Handbook of
MODERN GRINDING
TECHNOLOGY

OTHER OUTSTANDING VOLUMES IN THE CHAPMAN AND HALL
ADVANCED INDUSTRIAL TECHNOLOGY SERIES

Handbook of
MODERN GRINDING
TECHNOLOGY

Robert I. King
Robert S. Hahn

CHAPMAN AND HALL
NEW YORK   LONDON

First published 1986
by Chapman and Hall
29 West 35th St., New York, N.Y. 10001

Published in Great Britain by
Chapman and Hall Ltd
1 New Fetter Lane, London EC4P 4EE

© 1986 Chapman and Hall

Printed in the United States of America

Library of Congress Cataloging in Publication Data
King, Robert I. (Robert Ira), 1924-
Handbook of modern grinding technology.
(Chapman and Hall advanced industrial technology series)
Bibliography: p.
Includes index.
1. Grinding and polishing—Handbooks, manuals, etc.
I. Hahn, Robert S. II. Title. III. Series.
TJ1280.K53 1986   621.9′2   86-17626
ISBN 0-412-01081-X

Acknowledgments

We, the authors, acknowledge the excellent assistance given by the following organizations during the preparation of the text: Norton Company, 3M Company, Ohio State University, Pneumo Precision Company, and the Lockheed Missiles & Space Company. This text would not have been possible without their help.

The editors wish to give credit to both Kathleen Hahn, for her flawless typing and editing of the draft copies of this complex manuscript, and Donna King, for her support and suggestions during the development and integration of the text.

Dedication

This book on grinding is dedicated to all those who are searching for a way to improve the productivity of man and machine.

Contents

x   Contents

PREFACE

The latest information indicates that the United States now spends in excess of $150 billion annually to perform its metal removal tasks using conventional machining technology. That estimate is increased from $115 billion 5 years ago. It becomes clear that metal removal technology is a very important candidate for rigorous investigation looking toward improvement of productivity within the manufacturing system. To aid in that endeavor, an extensive program of research has developed within the industrial community with the express purpose of establishing a new scientific and applied base that will provide principles upon which new manufacturing decisions can be made.

One of the metal removal techniques that has the potential for great economic advantages is high-rate metal removal with related technologies. This text is concerned with the field of grinding as a subset of the general field of high-rate metal removal. Related processes (not covered in this text) include such topics as turning, drilling, and milling. In the final evaluation, the correct decision in the determination of a grinding process must necessarily include an understanding of the other methods of metal removal. The term grinding, as used herein, includes polishing, buffing, lapping, and honing as well as conventional definition: ". . . removing either metallic or other materials by the use of a solid grinding wheel."

The injection of new high-rate metal removal techniques into conventional production procedures, which have remained basically unchanged for a century, presents a formidable systems problem both technically and managerially. The proper solution requires a sophisticated, difficult process whereby management-worker relationships are reassessed, age-old machine designs reevaluated, and a new vista of product-process planning and design admitted. The key to maximum productivity is a "systems approach," even though a significant improvement in process can be made with the piecemeal application of good solid practice. This text was structured with those concepts in mind. However, the reader should also consider complementing subjects, such as machine dynamics, factory flow/loading, management psychology/strategy, and manufacturing economics. The "bottom line" is to increase the overall effectiveness of the factory from whatever devise that is reasonable, that is, to obtain the greatest return on the dollar invested.

As an example, consider the technical problem of increasing the speed of the grinding wheel. To realize the benefits of that increase, the table or spindle feedrate must be increased. That in turn has an impact on the basic machine design and the response of the control system. As the various speeds are increased, new dynamic ranges are encountered that could induce undesirable resonances in the machine and part being fabricated, requiring dampening consideration. The proper incorporation of an optimum grinding process into the factory requires the integration of all of the above technical considerations plus many others—a difficult systems solution requiring professional attention.

Finally, when making any major change in factory operations, the reader should consider the managerial style used. Keep in mind that the processes suggested in this text could deviate considerably from those that may exist in any particular factory environment. The use of new techniques would be ill advised if the operating employees are not supportive for any reason. Employee involvement and understanding during process change is necessary for success, and fear of the unknown is unacceptable.

Robert I. King
San Jose, California
Robert S. Hahn
Northboro, Massachusetts
February, 1986

Handbook of
MODERN GRINDING
TECHNOLOGY

CHAPTER 1

Part Processing by Grinding

Robert S. Hahn
Hahn Associates

Introduction

The problem of manufacturing high-quality parts at low cost confronts many companies. The cost of processing information required in the manufacturing process is substantial. With the introduction of computer-aided manufacturing, computer-aided process planning, and flexible manufacturing systems, the need for accurate, reliable process and equipment data is great. This chapter describes some of the factors involved in process planning, illustrates the detailed knowledge required to perform typical precision grinding operations, and describes computer-generated grinding cycles and multioperation grinding.

The Grinding Process-Planning Problem

The process-planning problem starts with the part print or data base prescribing the part dimensions, the tolerances, requirements for concentricity, roundness, cylindricity or flatness, squareness, surface finish, surface integrity, cycle time, and production requirements. Those "software data" or geometric part-print data must be processed to select the appropriate machine tool(s), grinding cycle parameters, and inspection equipment to produce "hard finished parts" satisfying all of the imposed specifications, as illustrated in Fig. 1.1. Each machine tool/equip-

3

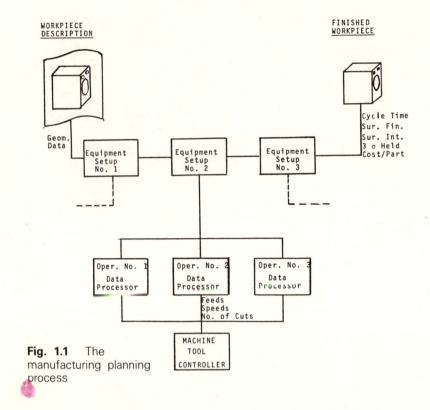

Fig. 1.1 The manufacturing planning process

ment must be set up. Then, on each setup, for example, Setup No. 2, 1 or more operations may be performed. Each operation, in turn, requires a data processor, which converts the geometric part-print specs into feeds and speeds. Those data must then be fed to the machine tool (grinder) controller. In the case of a typical grinding machine, the relationships between the machine characteristics, the grinding-process characteristics, and the machine input variables are indicated in Fig. 1.2. Accordingly, the task for the machine tool information processor (performed by humans or computer) is to generate the proper feeds and speeds for the grinding-machine controller based upon a knowledge of the machine-tool characteristics, Grinding-Process Characteristics, and Workpiece random variables, as illustrated in Fig. 1.2.

Grinding-Process Variables

In planning grinding operations it is necessary to define the various inputs and outputs and to develop relationships between them. In order

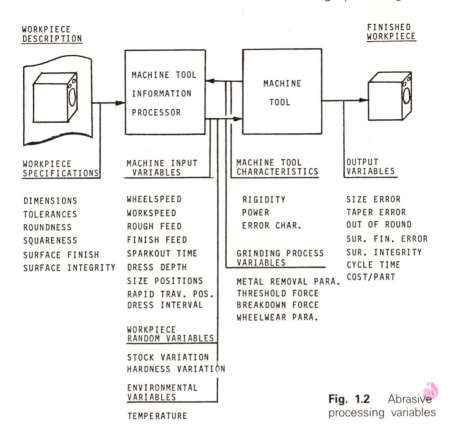

WORKPIECE
DESCRIPTION

FINISHED
WORKPIECE

MACHINE TOOL
INFORMATION
PROCESSOR

MACHINE
TOOL

WORKPIECE
SPECIFICATIONS

MACHINE INPUT
VARIABLES

MACHINE TOOL
CHARACTERISTICS

OUTPUT
VARIABLES

DIMENSIONS
TOLERANCES
ROUNDNESS
SQUARENESS
SURFACE FINISH
SURFACE INTEGRITY

WHEELSPEED
WORKSPEED
ROUGH FEED
FINISH FEED
SPARKOUT TIME
DRESS DEPTH
SIZE POSITIONS
RAPID TRAV. POS.
DRESS INTERVAL

RIGIDITY
POWER
ERROR CHAR.

GRINDING PROCESS
VARIABLES

METAL REMOVAL PARA.
THRESHOLD FORCE
BREAKDOWN FORCE
WHEELWEAR PARA.

SIZE ERROR
TAPER ERROR
OUT OF ROUND
SUR. FIN. ERROR
SUR. INTEGRITY
CYCLE TIME
COST/PART

WORKPIECE
RANDOM VARIABLES

STOCK VARIATION
HARDNESS VARIATION

ENVIRONMENTAL
VARIABLES

TEMPERATURE

Fig. 1.2 Abrasive processing variables

to do that, it is important to distinguish between input variables to the grinding machine and inputs to the grinding process, which occurs at the wheelwork interface. Typical inputs to grinding machines are: feed-rate or down feed, wheel- and workspeed, depth of dress, and sparkout time, as illustrated to Fig. 1.3. The input to the grinding process is the normal force developed at the wheelwork interface. The grinding process is quite analogous to the filing process as carried out by a machinist who places a bar of steel in a vise and files the end of the bar. The input to the filing process is the normal force exerted by the machinist pushing the file down against the work. The tangential force as well as the amount of chips removed and the resulting surface finish are outputs of the filing process. It is clear that the machinist cannot apply directly, a tangential force per se. Similarly, the normal force is the input to the grinding process while the tangential force, power, stock-removal rate, wheelwear rate, and surface finish are the output variables of the grinding process (see Chapter 2 for details).

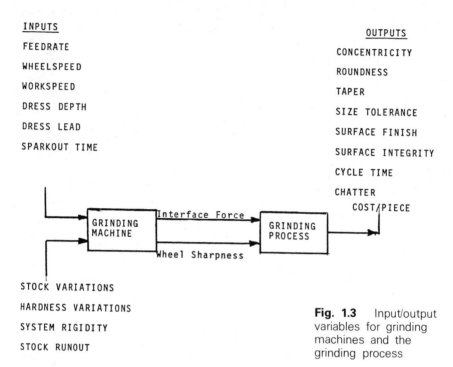

Fig. 1.3 Input/output variables for grinding machines and the grinding process

On conventional grinding machines the feedrate is controlled. As the grinding wheel engages the workpiece, forces are induced between wheel and work—the higher the force, the faster the stock removal. The induced force also governs the surface finish, the deflection in the machine, and the onset of thermal damage. Therefore, the induced force is one of the important variables that are uncontrolled in conventional feedrate-grinding machines.

The ability of the cutting surface of the grinding wheel to remove stock, called the wheel sharpness, is the second extremely important variable in the grinding process. In feedrate grinding, as the wheel sharpness drops (the wheel becomes dull or glazed), the induced force rises, resulting in increased deflection and, sometimes, thermal damage.

The size-holding ability of feedrate "sizematic" grinders is directly related to their ability to maintain the same force between wheel and work at the instant of retraction. If the induced force fluctuates in value at the termination point in the grinding cycle, the system deflection will also fluctuate and a size error will result unless in-process gaging is used. Even with in-process gaging, taper and surface finish fluctuations will occur. Variations in the induced force are caused by stock variations, wheel sharpness variations, workpiece hardness, and microstructure variations.

In cylindrical plunge-grinding operations, the interface force intensity (normal force per unit width of contact) is uniformly distributed over the face of the wheel. Accordingly, plunge grinding is the simplest type of grinding, as illustrated in Fig. 1.4. The feedrate $\bar{v}_f$ is applied to the cross slide of the machine. At the moment the wheel contacts the workpiece, the interface force intensity is zero. As the cross slide continues to move, the "springs" in the system compress, generating some interface force intensity F'_n. That causes the wheel and workpiece to mutually "machine" each other, the radius of the workpiece decreasing at the rate $\bar{v}_w$, the radius of the wheel decreasing at the rate $\bar{v}_s$, and the deflection x increasing at the rate $\dot{x}$, thus:

$$\bar{v}_w + \bar{v}_s + \dot{x} = \bar{v}_f \tag{1.1}$$

The feedrate of the cross slide $\bar{v}_f$ only equals the plunge-grinding velocity $\bar{v}_w$ when the wheelwear $\bar{v}_s$ is negligible and $\dot{x} = 0$ (the steady state).

The volumetric rates of stock removal Z'_w and wheelwear Z'_s per unit width of contact:

$$Z'_w = \pi \, D_w \, \bar{v}_w \tag{1.2}$$

$$Z'_s = \pi \, D_s \, \bar{v}_s \tag{1.3}$$

Fig. 1.4   The plunge grinding "mutual machining" of workpiece and grinding wheel

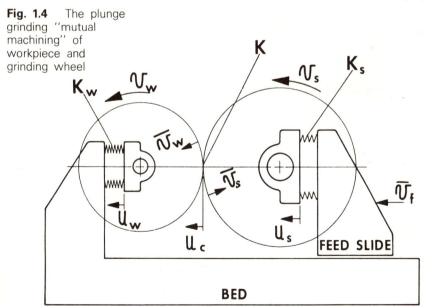

are plotted against the normal interface force intensity in Fig. 1.5, resulting in a "Wheelwork Characteristic Chart," which shows how a given wheelwork pair machine each other. The stockremoval curve Z'_w (solid dot) has a "rubbing region" at force intensities below F'_{th}, the "threshold force intensity," a "Plowing Region" for force intensities between F'_{th} and F'_{pc}, the "plowing-cutting transition," and a "cutting region" above F'_{pc}.[1,2,3] In the cutting region the abrasive grits remove chips in the usual way. In the plowing region they remove material by causing lateral plastic flow and highly extruded ridges to be formed along each side of the scratch, those ridges being removed by subsequent grits. The plowing region is important in obtaining good surface finishes. The cutting region is important in rounding up the workpiece and fast stock removal.

The slope of the Z'_w curve is called the "work removal parameter," WRP, or sometimes, λ_w, and is indicative of the "sharpness" S of the grinding wheel, defined as

$$S = \frac{(WRP)}{V_s} \qquad \left(\frac{m^2}{N}\right) \qquad (1.4)$$

which represents the cross-sectional area of a hypothetical ribbon of material being removed from the workpiece per unit normal force. The sharpness of grinding wheels is one of the most important variables in the grinding process and is frequently most difficult to control in practical grinding operations. Its value may vary 400% or 500%, causing size, taper, surface finish, and surface integrity problems.

The wheelwear curve Z'_s in Fig. 1.5 rises gradually at low-force intensity and then turns sharply upward (around 28 N/mm—150 lb./in.—in this case), at the so-called breakdown force intensity F'_{bd}. Precision grinding cycles must operate between F'_{th} and F'_{bd}. Curves for surface finish and power can also be shown on the wheelwork characteristic chart.

The "sharpness" S, or the "work removal parameter" WRP, for a given wheelspeed, wheelwork conformity, workpiece hardness and microstructure, is the direct indicator of the "real area of contact" in the wheelwork contact region and reflects changes in the real area of contact because of the development of wear flats on the abrasive grains. A typical variation of WRP with real area of contact is shown in Fig. 1.6. The WRP, therefore, provides a practical means for quantifying the sharpness of a grinding wheel during a grinding operation. (See Chapter 14, "Adaptive Control in Grinding.")

*Superscript numbers refer to papers listed in the bibliography

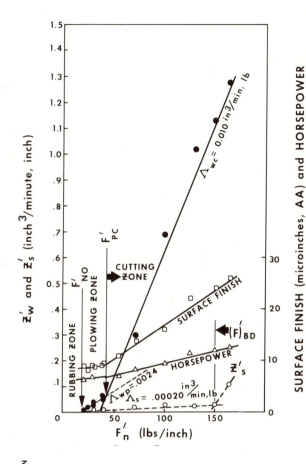

Fig. 1.5 Wheelwork characteristic chart showing the work-removal rate Z'_w, wheelwear rate Z'_s, surface finish, and power vs. interface normal force per unit width Wheel—A80K4 Dress Lead—.003 in/rev.; Dress Depth—.0002 in.; Wheelspeed—12000 SFPM; Workspeed—250 SFPM; Coolant—Cim cool 5 star; Work Material—AISI52100 @ 60 R_c; Equivalent Diameter—2.0 in.

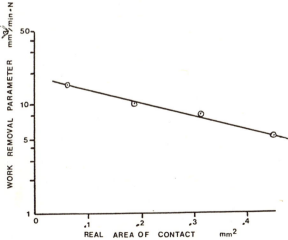

Fig. 1.6 Work-removal parameter vs. real area of contact (Ref. 4) Internal Grind Work Diameter—64mm; Workspeed—6m/s; Wheelspeed—30m/s; Constant Force Intensity—10n/mm; Material—AISI52100 60-62 R_c

Wheelwork Conformity

The difference in curvature of the wheel and work in the contact region has some effect on the cutting action at the wheelwork interface. The difference of curvature for internal or external grinding can be related to surface grinding by considering an "equivalent diameter" D_e of a surface grinding wheel having the same difference of curvature as the internal or external operation.

The "Equivalent Diameter" D_e illustrated in Fig. 1.7 is given by:

$$\frac{2}{D_e} = \Delta = \frac{2}{D_s} \mp \frac{2}{D_w} \tag{1.5}$$

$$D_e = \frac{D_w \ D_s}{D_w \pm D_s} \tag{1.6}$$

where the + or − sign is used for external or internal grinding With that parameter, internal, external, and surface grinding can be related.

In grinding shoulders with wheels inclined at the angle β, the radii of curvature of the abrasive wheel in the X and Z directions shown in Fig. 1.8 are:

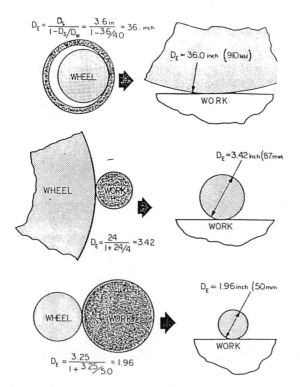

$$D_E = \frac{D_s}{1 - D_s/D_w} = \frac{3.6 \text{ in}}{1 - 3.6/40} = 36. \text{ inch}$$

$D_E = 36.0 \text{ inch } (910 \text{ MM})$

$$D_E = \frac{24}{1 + 24/4} = 3.42$$

$D_E = 3.42 \text{ inch } (87 \text{ mm})$

$$D_E = \frac{3.25}{1 + 3.25/5.0} = 1.96$$

$D_E = 1.96 \text{ inch } (50 \text{ mm})$

Fig. 1.7   Relating the difference of curvature between wheel and workpiece for internal or external grinding to the "equivalent diameter" of a surface grinder wheel.

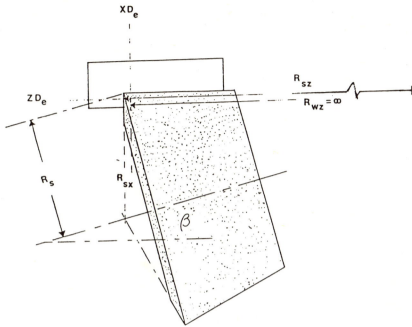

Fig. 1.8   Illustration of the "equivalent diameter" in the radial direction (XD_e) and the axial direction (ZD_e) for vector grinding of a shoulder and adjacent diameter. R_{sz} = Radius of curvature of wheel in axial direction; $R_{wz} = \infty$ = Radius of curvature of shoulder; R_{sx} = Radius of curvature of wheel in radial direction; R_s = Radius of wheel

$$R_{sx} = \frac{D_s}{2 \cos\beta} \tag{1.7}$$

$$R_{sz} = \frac{D_s}{2 \sin\beta} \tag{1.8}$$

so that:

$$XD_e = \frac{D_w \dfrac{(D_s)}{(\cos\beta)}}{D_w \pm \dfrac{(D_s)}{(\cos\beta)}} \tag{1.9}$$

$$ZD_e = \frac{D_s}{\sin\beta} \tag{1.10}$$

The large ZD_e on the shoulder is conducive to thermal damage and reduced WRP.

Basic Plunge-Grinding Relations for Conventional Cylindrical Grinding*

Stock removal, wheelwear, surface finish, and power and force relationships can be developed from the wheelwork characteristic chart illustrated in Fig. 1.5. With the plowing region in Fig. 1.5 neglected for simplicity, the stock-removal relation is:

$$Z'_w = WRP (F'_n - F'_{th}) \text{ or:} \tag{1.11}$$

$$\pi D_w \bar{v}_w = WRP (F'_n - F'_{th}) \text{ or:} \tag{1.12}$$

$$\bar{v}_w = \frac{WRP (F'_n - F'_{th})}{\pi D_w} \tag{1.13}$$

Eq. 1.13 gives the plunge-grinding velocity in terms of the normal interface force intensity.

When the wheelwear rate is negligible ($\bar{v}_s < \bar{v}_w$) and a steady state of deflection exists ($\dot{x} = 0$). Eq. 1.1 becomes:

$$\bar{v}_w = \bar{v}_f \tag{1.14}$$

Eq. 1.13 can be solved for F'_n with $\bar{v}_w$ replaced by $\bar{v}_f$, thus

$$F'_n = \frac{\pi D_w \bar{v}_f}{WRP} + F'_{th} \tag{1.15}$$

That gives the induced force intensity generated by the feed rate $\bar{v}_f$ in the steady state and provides a basic understanding of feedrate grinding.

It will be seen that, at a given feedrate, the induced force increases as the wheel becomes dull. That is caused by (1) a drop in WRP and (2) an increase in threshold force F'_{th}.

The wheelwear curve Z'_s in Fig. 1.5 below F'_{bd}, may be approximated according to Lindsay[2] by:

$$Z'_s = WWP (F'_n)^2 \tag{1.16}$$

When the wheelwear parameter, WWP, is known, wheelwear can be estimated for various grinding conditions.

The breakdown force intensity F'_{bd} in Fig. 1.5 may be estimated for Al_2O_3 vitrified wheels[3] by:

$$F'_{bd} = 62.3 (vol)^{.55}(D_e)^{.25} (N/cm) \tag{1.17}$$

---

*Primed quantities signify per unit width

Precision grinding cycles should be designed so that the induced force intensity lies below F'_{bd}.

The wheel depth-of-cut h (advance of wheel per work revolution) is given by:

$$h = \frac{\bar{v}_w}{N_w} \quad \text{(um)} \tag{1.18}$$

That relation permits all the results developed for cylindrical grinding to be applied to surface grinding operations at normal workspeeds.

The "work cutting stiffness" K_c (normal force required to take unit depth of cut) is an important quantity governing the rate of rounding up, the sparkout time, and chatter behavior when compared with the "system stiffness" K_m. It is given by:

$$K_c = \frac{F_n}{h} = \frac{V_w W}{WRP} \quad \text{(N/mm)} \tag{1.19}$$

The dimensionless "machining-elasticity number," α, formed by the ratio:

$$\frac{K_c}{K_m} = \alpha \tag{1.20}$$

relates elastic effects in machining or grinding operations to the stiffness of the machine tool.

The power P absorbed in the grinding process is;

$$P = F_t V_s \quad \left(\frac{Nm}{sec} \text{ or watts}\right) \tag{1.21}$$

The ratio F_t/F_n varies between .3 for a dull wheel and .7 for a sharp wheel with an average value of .5. Therefore:

$$F_t \cong .5 F_n \tag{1.22}$$

and:

$$P \cong \tfrac{1}{2} F_n V_s \tag{1.23}$$

Using Eq. 1.15 and neglecting the threshold force F_{th},

$$P = \frac{\pi D_w W \, V_s}{2 \, WRP} \, \bar{v}_f \tag{1.24}$$

gives the power required for any feedrate $\bar{v}_f$.

The "specific power," P_s, using Eqs. 1.11, 1.15 and neglecting F'_{th} is:

$$P_s = \frac{P'}{Z'_w} = \frac{V_s}{2 \, WRP} \tag{1.25}$$

and is inversely proportional to the WRP.

The "time constant," τ_0, of a grinding system governs the time required to build up grinding force or sparkout. It depends upon the system stiffness K_m and on the material being ground.[3]

$$\tau_0 = \frac{\pi D_w W}{WRP \ K_m} \quad (sec) \tag{1.26}$$

On materials exhibiting a plowing region, the time constant suddenly changes during a sparkout when the plowing region is encountered. The WRP in the plowing region is about ⅓ WRP in the cutting region.

The "G Ratio," giving the ratio of the volume of metal removed to the volume of abrasive consumed, is generally a variable depending upon the particular operating force intensity. It is a valid ratio only in the case where the Z'_w vs F'_n and the Z'_s vs F'_n relations are straight lines emanating from the origin in Fig. 1.5. Generally, that is not true, and, accordingly, the G ratio will vary as the feedrate is changed.

The above relations will be used later for calculating grinding cycle parameters. In the next section various grinding configurations are considered.

Grinding Configurations

Multioperation Grinding

Many workpieces often require a number of surfaces to be ground on each individual workpiece. If several of those surfaces can be ground in 1 operation, production efficiences can often be achieved. If they cannot all be ground in 1 operation, then several operations are required. However, production efficiencies can also be achieved if multiple operations can be performed for 1 staging of the workpiece, thereby reducing the number of additional setups, part handling, and ensuring squareness and concentricity.

Figure 1.9 illustrates, for example, a CNC grinder with 2 wheelheads mounted on the same cross-slide, the left-hand head grinding 6 surfaces in operation 4 and the right-hand head subsequently grinding 3 surfaces in operations 1, 2, and 3. Multiple grinding operations can sometimes also be performed by using "compound wheels," where several cutting surfaces, or wheels, are provided on the same wheelhead to execute several operations, as illustrated in Fig. 1.10. An axial feed (ZFEED) is used to grind operation 1, while a vector feed (where computer control provides simultaneous feedrates to both X and Z axes) subsequently executes operation 2 with the smaller outboard section of the grinding

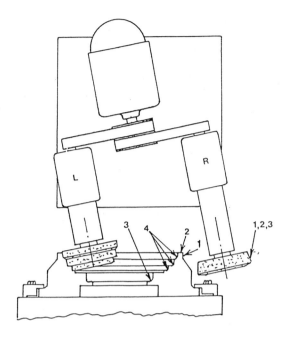

Fig. 1.9
Multioperation grinding
of 10 surfaces using 2
wheelheads on the
same cross-slide to
perform operations 1,
2, and 3 with the
right-hand head and
operation 4 with the
left-hand head.

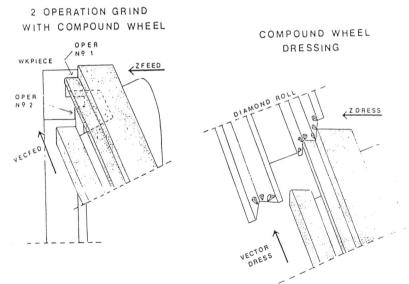

Fig. 1.10   Illustration of a diamond roll dressed compound wheel for
executing 2 operations with different areas on the wheel.

wheel. Also shown is the diamond dress roll for providing compound wheel dressing.

With computer control over both cross-slide and axial motions (X and Z axes), external or internal grinding of tapered parts can be performed as illustrated in Fig. 1.11b. In that example the tapered section A of the wheel is used to grind the taper A, where the X and Z slides under computer control perform a vector reciprocation along the surface A. Following that operation, the straight section B is ground, and, finally, the shallow taper C is ground.

Multiplunge Roughing with Reciprocating Finish Grind

Fig. 1.11a illustrates a simple straight bore to be ground. In conventional reciprocate grinding the wheel is usually fed in manually to contact the work, develop a spark stream, and then fed axially to traverse the length of the bore. If there is more stock in another section of the bore, excessive forces are developed as the wheel traverses along the bore. Generally, those forces cause either the leading edge or the trailing edge of the wheel, depending upon the angular stiffness of the wheel spindle, to break down. If the angular stiffness is high, the leading edge tends to break down. If it is low, the leading edge tends to deflect out of the cut, concentrating the grinding force on the trailing edge and causing it to break down. In either case the wheel face is not being used effectively. With CNC grinding, those difficulties can be overcome by making a series of adjacently spaced plunge grinds to rough grind the part, correcting tapered stock conditions, and bringing the part to so-called first size, and then to finish grind to size by reciprocating the wheel in the usual manner. That method avoids wheel breakdown during the rough grind but still uses the wheel aggressively to remove stock.

OD/ID Grinds with Adjacent Shoulders

A number of methods of grinding OD or ID parts with adjacent shoulders are illustrated in Fig. 1.12. Their advantages and disadvantages are discussed below.

Fig. 1.12a shows a wheel making a plunge grind in the X direction, where it must feed the entire width of the shoulder before it strikes the OD/ID stock (XSTKAL). It is a very time-consuming operation and causes considerable wheelwear on the left-hand edge of the wheel. A somewhat improved situation occurs in Fig. 1.12b, where the wheel is rapidly traversed into the corner and then fed on a vector to grind the

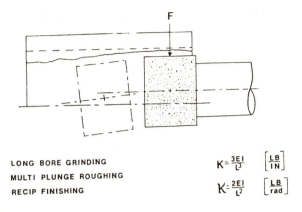

LONG BORE GRINDING

MULTI PLUNGE ROUGHING

RECIP FINISHING

$$K = \frac{3EI}{L^3} \quad \left[\frac{LB}{IN}\right]$$

$$\kappa = \frac{2EI}{L^2} \quad \left[\frac{LB}{rad}\right]$$

a

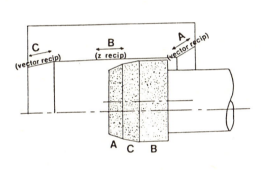

b

Fig. 1.11 Long bore or OD traverse grinding. a) illustrating the lateral and angular deflection of a cantilevered wheel for multiplunge roughing followed by reciprocating finish grinding where K is the lateral stiffness and κ is the angular stiffness. b) illustrating vector reciprocation to grind 3 surfaces.

shoulder and the OD/ID simultaneously. If the vector angle is chosen so that:

$$\tan\theta = \frac{XSTKAL}{ZSTKAL} \tag{1.27}$$

size on the shoulder will be reached at the same time as size on the OD/ID. The disadvantage of that method is that the shoulder may be "burned," cracked, or thermally damaged metallurgically because of excessive heat.

A cooler grind on the shoulder can be obtained by tilting the wheelhead axis slightly, as shown in Fig. 1.12c, where the end cutting face of the wheel now has a curvature in the Z direction, thereby increasing the difference in curvature between wheel face and shoulder.

Fig. 1.12d, e, and f illustrate similar situations where a "back face"

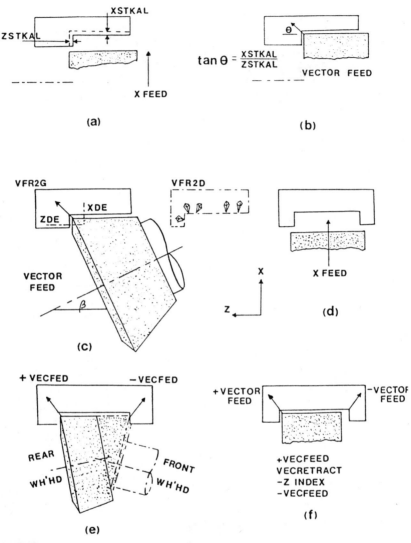

Fig. 1.12   Several methods of grinding shoulders and adjacent diameters

must be ground. Fig. 1.12e represents the coolest grind but requires 2 wheelheads on the same cross slide.

A shoulder and adjacent OD/ID can also be ground with only 1 axis of motion under feed by tipping the work axis relative to the feed direction, as illustrated in Fig. 1.13a for an ID grind and 1.13b for an OD grind.

In those cases the shoulder dimension and the OD/ID dimension are

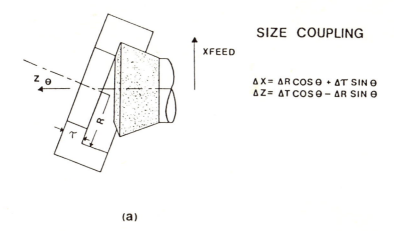

SIZE COUPLING

$\Delta X = \Delta R \cos\theta + \Delta T \sin\theta$
$\Delta Z = \Delta T \cos\theta - \Delta R \sin\theta$

XFEED

(a)

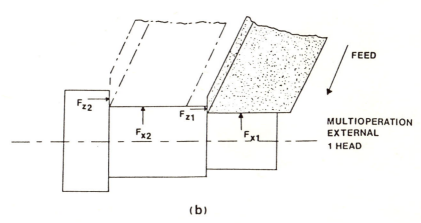

FEED

MULTIOPERATION
EXTERNAL
1 HEAD

(b)

Fig. 1.13 Examples of the size-coupling effect where changes in diameter cause changes in shoulder dimension

coupled through the size-coupling effect, where a change in one dimension will cause an (unwanted) change in the other dimension. Readjustment of the Z position can be made to compensate for a change in radial dimension according to:

$$\Delta X = \Delta R\cos\theta + \Delta\tau\sin\theta \qquad (1.28)$$

$$\Delta Z = \Delta\tau\cos\theta - \Delta R\sin\theta \qquad (1.29)$$

By setting $\Delta\tau = 0$, (the condition that the shoulder dimension does not change), values of ΔX and ΔZ are obtained that permit readjustment of the crossslide and table to effect a change in diameter size only.

Form Grinding

In addition to grinding shoulders adjacent to an OD/ID, some parts often require form-grinding operations. Those parts sometimes exhibit problems in thermal damage, local variations in surface finish, or loss of profile accuracy. In dealing with those problems, it is helpful to understand the grinding process variables that cause them. Those variables are discussed below.

The plunge grinding of a ball track in a ball-bearing race can be used to illustrate the effects found in many form-grinding operations, as shown in Fig. 1.14. There are 4 local process variables, which vary from point to point along the profile.

Fig. 1.14a illustrates the distribution of the "Equivalent Diameter D_e along the profile. The "Equivalent Diameter" is a measure of the difference in curvature between the abrasive wheel and the workpiece in the contact zone.

Fig. 1.14b illustrates the distribution of the local normal-force intensity (normal force per unit width of cut) along the profile, while the dashed line represents the distribution of the local "threshold force intensity." The length of the wheelwork contact zone L_c is also shown. The large value of equivalent diameter, D_e, on the side walls, the long length of contact L_c, and the normal-force intensity often combine to exceed the threshold conditions for thermal damage, resulting in burn on the sidewalls. Surface finish is generally better at the bottom of the ball track, where the induced force intensity is low.

Computer-Aided Process Planning

Based upon the equations developed in this chapter and in Chapter 2 the grinding performance of various configurations can be predicted by computer simulation. Grinding-cycle parameters can be generated in terms of geometrical input data. Those programs are predicated on the work removal parameter, WRP, and threshold force F_{th}, which, essentially, describe the combination of workpiece machinability and degree of wheel sharpness for a given wheelwork pair.

The work removal parameter may, in some instances, be estimated by equations given in Chapter 2. A discussion of machinability parameters in grinding has also been given by Hahn.[5,6] In addition, there are methods for measuring WRP or, λ_w, and threshold force for wheelwork pairs

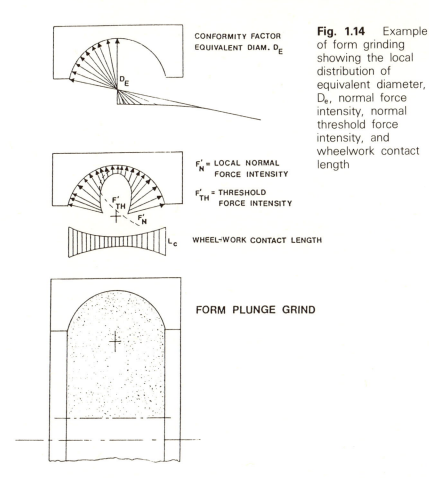

CONFORMITY FACTOR
EQUIVALENT DIAM. D_E

F'_N = LOCAL NORMAL
       FORCE INTENSITY

F'_{TH} = THRESHOLD
        FORCE INTENSITY

L_c WHEEL-WORK CONTACT LENGTH

FORM PLUNGE GRIND

Fig. 1.14 Example of form grinding showing the local distribution of equivalent diameter, D_e, normal force intensity, normal threshold force intensity, and wheelwork contact length

on machines currently in use. In any event, once accurate values of WRP and F_{th} have been acquired, reasonably accurate cycle parameters and output grinding performance data can be obtained.

Lindsay[7] has measured the WRP and F_{th} for several wheels grinding AISI52100 and M50 steels under a variety of conditions. Fig. 1.15 gives the values of WRP (or λ_w) as a function of wheel surface speed and equivalent diameter D_e. Fig. 1.16 gives similar data for M50, while Fig. 1.17 gives threshold force intensities for both AISI52100 and M50. Using appropriate values from those figures as input to the computer program, one can compare the grinding performance when internally grinding a 24mm hole in AISI52100 and M50 steel.[6,8]In that case there is a "race" to round up the rough stock before it is consumed, and the computer program will reduce the feedrate, if necessary, by a certain percentage in order to have enough time to round up. That is shown in Table 1—1 as

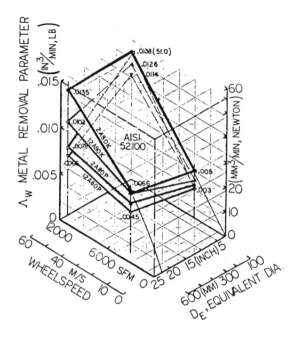

Fig. 1.15 Variation of the work-removal parameter with wheelspeed and equivalent diameter for 4 wheel grades; 2A80K, 12A80K, 2A80P, 12A80P, Grinding AISI52100 at $R_c = 63$ and "Flowrex 100" coolant. (after Lindsay (7))

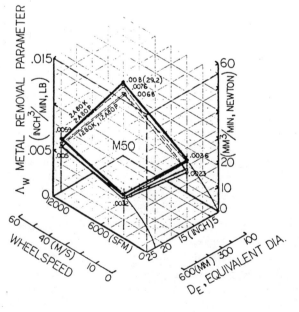

Fig. 1.16 Variation of the work-removal parameter with wheelspeed and equivalent diameter for 4 wheel grades; 2A80K, 12A80K, 2A80P, and 12A80P Grinding M50 at $R_c = 63$ and "Flowrex 100" coolant. (after Lindsay (7))

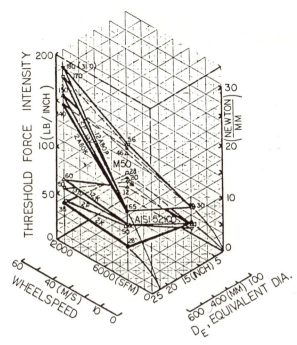

Fig. 1.17 Variation of the threshold force intensity with wheelspeed and equivalent diameter for 4 wheel grades; 2A80K, 12A80K, 2A80P and 12A80P Grinding AISI52100 and M50 at $R_c = 63$ and "Flowrex 100" coolant. (after Lindsay (7))

the "percent reduction owing to roundup constraint" along with the other grinding performance outputs. Table 1—2 gives the input parameters. Note that the threshold force intensity (F'_{no}) for the "new" large wheel grinding AISI52100 is 4 times as large as that for the "used" wheel because of the larger D_e. That causes the "new" wheel to cut less freely and thus a 70% reduction in feedrate or force from the breakdown value F'_{bd} of the new wheel. Also, an "active cycle time" (excluding load and unload time and time for machine movements) of 5 to 6 sec. is appropriate for a 25mm bore with .76mm stock on diameter.

For the M50 material, there is a great difference in active cycle time for the "new" and "used" wheels. That is again caused by the much larger threshold force intensity (F'_{no}) characteristic of M50 at large values of D_e (31.3 N/mm vs. 8.74 N/mm for AISI52100). The result is a 29.39 sec. active grind time. The rough grind feedrate is also much lower because of the large threshold force, which leaves a relatively small amount of force for stock removal.

Other types of grinding configurations can also be simulated by computer. For example, consider the long bore or OD grind in Fig. 1.11a. As mentioned earlier, reciprocate grinding under heavy forces (rough grind) is not so efficient as multiplunge roughing followed by reciprocate finish grinding. Table 1—4 gives the grinding parameters and out-

Table 1—1   Grinding cycle performance

Material AISI52100	
Used Wheel 17.78mm	
Stock Required for Roundup	.439mm
Roundup Time	3.06sec
Rough Grind Time	3.36sec
Dress Time	.67sec
Sparkout Time	2.00sec
Active Cycle Time	6.03sec
Rough Grind Feedrate	.130mm/sec
% Reduction owing to Roundup Constraint	20%
New Wheel 22.22mm	
Stock Required for Roundup	.421mm
Roundup Time	3.12sec
Rough Grind Time	3.54sec
Dress Time	.67sec
Sparkout Time	.80sec
Active Cycle Time	5.01sec
Rough Grind Feedrate	.123mm/sec
% Reduction owing to Roundup Constraint	70%
Material M50	
Used Wheel 17.78mm	
Stock Required for Roundup	.456mm
Roundup Time	5.82sec
Rough Grind Time	6.24sec
Dress Time	.67sec
Sparkout Time	2.10sec
Active Cycle Time	9.01sec
Rough Grind Feedrate	.0711mm/sec
% Reduction owing to Roundup Constraint	40%
New Wheel 22.22mm	
Stock Required for Roundup	.254mm
Roundup Time	19.56sec
Rough Grind Time	28.62sec
Dress Time	.67sec
Sparkout Time	.10sec
Active Cycle Time	29.39sec
Rough Grind Feedrate	.0137mm/sec
% Reduction owing to Roundup Constraint	0%

Table 1—2   Grinding input data

New Wheel Diameter	22.22mm
Used Wheel Diameter	17.78mm
Grit Size	80
Wheel Hardness	K
Wheel Structure	6
Bore Diameter	24.13mm
Length of Bore	25.4mm
Rockwell Hardness	63 C
Desired Surface Finish	.375um
Max. Stock	.813mm
Min. Stock	.711mm
Rough Stock Runout (TIR)	.254mm
Final Runout	5.0um
Workspeed	1000 rpm
Wheelspeed	30000 rpm
Quill Diameter	17.53mm
Quill Length	50.mm
AISI52100	
A_w New Wheel	.406mm³/sec.N
A_w Used Wheel	.701mm³/sec.N
F'_{NO} New Wheel	8.74 N/mm
F'_{NO} Used Wheel	2.10 N/mm
F'_{BD} New Wheel	48.3 N/mm
F'_{BD} Used Wheel	33.7 N/mm
M50	
A_w New Wheel	.332mm³/sec.N
A_w Used Wheel	.369mm³/sec.N
F'_{NO} New Wheel	31.3 N/mm
F'_{NO} Used Wheel	3.5 N/mm
F'_{BD} New Wheel	48.4 N/mm
F'_{BD} Used Wheel	33.7 N/mm

Table 1—3   Grinding input data

Rear Wheelhead Data, Part No. 123456	
No. of Operations	+000001
Coolant? (Emul.=0,	
Chem. Sol.=1,	
Oil=2)	+000000
Auto Load=1, Hand Load=0	+000000
Workspeed (rpm)	+000300
Work Stall Torque (N*M)	+000057
Wheelspeed (rpm)	+025000
New Wheel Z Depth of Dress (um)	+000000
New Wheel X Depth of Dress (um)	+000010
Width of Wheel at Cutting Surface (mm)	+000040
Width of End Cutting Face (mm)	+000000
New Wheel Diameter (mm)	+000064
Used Wheel Diameter (mm)	+000050
Abrasive Grit Size	+000080
Wheelhead Swivel Angle (deg.)	+000000
Wheelhead Power (kW)	+000020
Wheel Structure No.	+000005
Wheel Hardness (H,I,J,—=0,1,2,—)	+000000
Surface Finish on Bore (100*RA) (um)	+000080
X Stock Allowance (um)	+000300
Length of Bore (um)	+125000
Recip. Bore Stroke (um) (after first size)	+100000
Work Diameter (um) (± for OD/ID)	−075000

Table 1—4   Grinding output data

Data for Used Wheel D_s = 50mm	
Z Slide Traverse Rate (mm/min)	6000.0000
Z Slide Traverse Rate (mm/min)	1900.0000
X Equivalent Diameter (cm)	15.0000
X Wheel Depth of Cut (roughing)	9.9467
X Wheel Depth of Cut (finishing)	4.9299
X Breakdown Force (dan) (dekanewtons)	93.5261
X Grinding Force (n)	427.9104
X Compensation for Wheelwear (um)	43.5629
X Rough Feedrate (um/sec)	49.7335
Finish Feedrate (um/sec)	24.6493
Finish Stock on Radius (um)	103.3786
Threshold Force X Direction (n)	0.0000
Sparkout Time (sec)	2.5422
Active Grind Time (Feedrate) (sec)	44.5428
Active Grind Time (Force Adaptive) (sec)	22.1696
Data for New Wheel D_s = 64mm	
Z Slide Traverse Rate (mm/min)	6000.0000
Z Slide Traverse Rate (mm/min)	1900.0000
X Equivalent Diameter (cm)	43.6364
X Wheel Depth of Cut (roughing)	10.0585
X Wheel Depth of Cut (finishing)	9.4988
X Breakdown Force (dan) (dekanewtons)	122.1440
X Grinding Force (n)	427.9104
X Compensation for Wheelwear (um)	33.2978
X Rough Feedrate (um/sec)	50.2923
Finish Feedrate (um/sec)	47.4940
Finish Stock on Radius (um)	154.8690
Threshold Force X Direction (n)	0.0000
Sparkout Time (sec)	2.4036
Active Grind Time (Feedrate) (sec)	35.5044
Active Grind Time (Force Adaptive) (sec)	17.5519

put performance for both reciprocate roughing and multiplunge roughing of a 75mm hole, 125mm long with .3mm radial stock, using a 64mm diameter "new" wheel and a 50mm diameter "used" wheel. Table 1—3 gives the input data.

The program seeks the highest force, or rough feedrate, that will not stall the wheelhead and the workhead or break the wheel down. Note that the program gives the amount of wheelwear to be expected, the finish feedrate to achieve the desired surface finish, and the active grind time for reciprocate roughing (feedrate grinding) and for the multi-plunge roughing method (force adaptive).

In concluding this chapter, it should be noted that the acquisition of successful grinding cycle parameters and satisfactory output perform-ance is not always easily achieved; but with the continued development of grinding parameter data bases and relationships (see Chapter 2), cou-pled with adaptive control techniques (see Chapter 14), improved grind-ing performance and reliability may be expected.

Nomenclature

V_w Work Surface Speed (m/sec)
N_w Workspeed (RPM)
V_s Wheel Surface Speed (m/sec)
N_s Wheelspeed (RPM)
D_w Work Diameter (mm)
D_s Wheel Diameter (mm)
ℓ Dress Lead (um/rev)
c 2*diamond depth-of-dress (um)
d Grain Diameter (um)
D_e Equivalent Diameter (cm)
$vol = 1.33H + 2.2S - 8$ (vol % of bond in wheel)
        $H = 0,1,2,3$—for H,I,J,K—hardness
        S = Wheel Structure Number
F_n' Normal Force Intensity (N/mm)
F_n Normal Force (N)
F_t Tangential Force (N)
F_{th}' F_{no}' = Threshold Force Intensity (N/mm)
F_{pc}' Plowing-Cutting Transition Force Intensity (N/mm)
F_{bd}' Wheel Breakdown Force Intensity (N/mm)
h Wheel Depth-of-Cut (um)
p Grinding Power (watts)
P_s Specific Power (joules/cu mm)
τ_0 Time Constant (sec)

R_c Rockwell Hardness C Scale

$WRP = \Lambda_w$ = Work Removal Parameter (cu mm/min∗N)

K_m System Stiffness (N/um)

W   Width of Wheelwork Contact (mm)

$\bar{v}_f$ Feedrate (um/sec)

$\bar{v}_w$ Penetration Velocity of Wheel into Work (um/sec)

$\bar{v}_s$ Radial Wheelwear Velocity (um/sec)

Z_w Volumetric Rate of Stock Removal (cu mm/sec)

Z'_w Stock Removal Rate per Unit Width (sq mm/sec)

Z_s Volumetric Wheelwear Rate (cu mm/sec)

Z'_s Wheelwear Rate per Unit Width (sq mm/sec)

S   Wheel Sharpness (sq m/N)

WWP   Wheelwear Parameter (mm^4/min∗N^2)

References

1.   R. S. Hahn, *On the Nature of the Grinding Process*, Proceedings of the Third Mtdr Conf., 1961, pp. 129–54, Permagon Press, 1961.

2.   R. P. Lindsay, "On Metal Removal and Wheel Removal Parameters, Surface Finish, Geometry and Thermal Damage in Precision Grinding," Ph.D. Dissertation, Worchester Polytechnic Institute, 1971.

3.   R. S. Hahn, R. P. Lindsay, "Principles of Grinding," *Machinery*, 1971. July, August, September, October, November, 1971.

4.   R. S. Hahn, R. P. Lindsay "On the Effects of Real Area of Contact and Normal Stress In Grinding," *Annal of C.I.R.P.*, Vol. XV, 1967, pp. 197–204.

5.   R. S. Hahn, "A Survey on Precision Grinding for Improved Product Quality," Proc. 25th MATADOR Conference, Birmingham, England, 1985.

6.   R. S. Hahn, "The Influence of Grinding Machinability Parameters on the Selection and Performance of Precision Grinding Cycles," Proc. ASM/SME International Conference on Machinability Testing and Utilization of Machining Data, Oak Brook, Ill. Sept. 12–13, 1978.

7.   R. P. Lindsay, "Variables Affecting Metal Removal and Specific Power in Precision Grinding," SME Paper MR71–269, SME, Dearborn, Mich. 48128.

8.   R. S. Hahn, *Grinding Software–An Important Factor in Reducing Grinding Costs*, Proc. Int. Conf. on Production Engineering, Japan Soc. of Prec. Engineering, Tokyo, 1974.

CHAPTER 2

Principles of Grinding

Richard P. Lindsay
Norton Company

Introduction

There are many types of grinding: "precision" and "rough," internal, external, surface, centerless; using wheels or belts; and conventional or "super" abrasives. I believe they all act the same. When a moving abrasive surface contacts a workpiece, if the force is high enough, material will be removed from the part and the abrasive surface will wear. Those two things will always occur; however, the force level determines how fast the mutual removal rates will be, how rough the remaining surface will be, and whether the workpiece will be metallurgically damaged or not. The purpose of this section is to provide relationships between variables and to illustrate how changes to a system affect its performance.

Basic System Features

Fig. 2.1 illustrates an external grinding system, but the discussion will pertain to any configuration. The subscript "W" will be used for all workpiece terms and the subscript "S" for the wheel. (This is international practice and is used by most writers outside the United States.) Thus, Fig. 2.1 and the table below describe various parameters.

The width of the wheelwork contact is B. It is more convenient to use volumetric removal rates than the radial rates. That enables systems of different sizes to be related to each other. Volumetric removal rates have

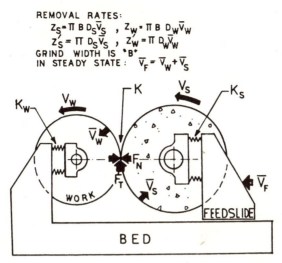

REMOVAL RATES:
$$Z_S = \pi\, B\, D_S \bar{V}_S \quad,\quad Z_W = \pi\, B\, D_W \bar{V}_W$$
$$Z'_S = \pi\, D_S \bar{V}_S \quad,\quad Z'_W = \pi\, D_W \bar{V}_W$$
GRIND WIDTH IS "B"
IN STEADY STATE : $\bar{V}_F = \bar{V}_W + \bar{V}_S$

Fig. 2.1   Cylindrical grinding system schematic

been labeled "Z" (and, more recently, "Q") in international publications.

There are 2 forces: F_n normal to the wheelwork contact surface and F_t tangential to the contact surface. Multiplying F_t by wheelspeed gives the power used in grinding (when an appropriate constant relating force and speed to power is used). International convention uses "primed" quantities to illustrate factors "per unit width." Thus, volumetric removal rates and forces per unit width would be written as: F'_n, F'_t, Z'_w, Z'_s (or Q'_w and Q'_s).

These conventions are confusing at first, but once a reader gets comfortable with them, he can read papers from around the world easily without continuously checking what the symbols mean.

Fig. 2.2 shows a surface grinding configuration in order to illustrate

Table 2—1

	Rotational Speed	Diam.	Radial Removal Rate	Total Volumetric Removal Rate: Z	Volumetric Removal Rate per Unit Width: Z'
Work	V_w	D_w	$\bar{v}_w$	$\pi D_w \bar{v}_w B$	$\pi D_w \bar{v}_w$
Wheel	V_s	D_s	$\bar{v}_s$	$\pi D_s \bar{v}_s B$	$\pi D_s \bar{v}_s$
Units	ft/minute	inch	inch/min	inch³/min	inch³/min, in

GRIND WIDTH IS "B"
REMOVAL RATES : $Z_W = a V_W B$
$Z'_W = a V_W$

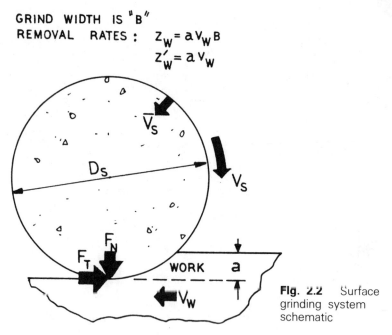

Fig. 2.2   Surface grinding system schematic

another way of calculating volumetric removal rates. Here the depth of cut is "a," and the width of the contact is B. The volumetric removal rate here is:

$$Z = aV_wB \text{ (inch}^3/\text{min)} = \text{(inch) (inch/min) (inch)} \qquad (2.1)$$

and the removal rate per unit width is:

$$Z'_w = aV_w \text{ (inch}^3/\text{min, inch)} = \text{(inch) (inch/min)} \qquad (2.2)$$

The concept of "depth of cut" can be applied to cylindrical grinding, too. In Fig. 2.1, if the workpiece rotational speed is N_w (revolutions/minute), then the depth of cut

$$a = \frac{\bar{v}_w}{N_w} \quad \frac{\text{(inch)}}{\text{(rev.)}} = \frac{\text{inch/min}}{\text{rev/min}} \qquad (2.3)$$

and the volumetric removal rate can also be calculated as:

$$Z_w = a BV_w \qquad (2.4)$$

Substituting Eq. (2.3) into (2.4):

$$Z_w = \frac{v_w}{N_w} BV_w \quad \text{and noting that } \pi D_w N_w = V_w, \text{ then}$$

$$Z_w = \frac{\bar{v}_w}{N_w} B(\pi D_w N_w), \text{ or}$$

$$Z_w = \pi D_w \bar{v}_w B, \text{ which is the same as given in Table 2-1}$$

In Fig. 2.1, $\bar{v}_f$ is the infeed rate of the feed slide. When the wheel contacts the work, the continuous infeed motion compresses the springs (K_w and K_s) building up the normal force F_n. That causes the wheel to wear at the $\bar{v}_s$ rate and the workpiece to be ground at the rate $\bar{v}_w$. If the steady rate is reached, F_n and F_t are constant and $\bar{v}_w + \bar{v}_s = \bar{v}_f$. If later in the cycle a second slower finish-grind rate $\bar{v}_f$ is programmed, the system springs will relax to generate a new lower force, causing slower grinding and wheelwear rates, $\bar{v}_w$ and $\bar{v}_s$. In both conditions, however, the infeed rate compresses the springs, generating the normal force that becomes the independent input to the grinding system. It is F_n that causes $\bar{v}_w$ and $\bar{v}_s$, not the infeed rate of the slide.

System Relationships: Easy-to-grind-Steels

System Characteristic Chart

Figure 2.3 shows how Z'_w, Z'_s Power and surface finish are related to normal force, F'_n. The caption shows those results were for grinding 52100 (R_c60) steel at a wheelspeed of 10,300 fpm. The width of grind was 0.375 inch. (The factor "D_e" will be discussed later.) Notice that below 20

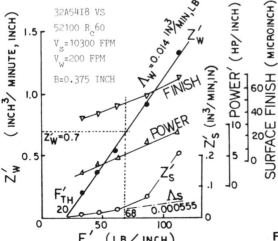

Fig. 2.3   The system characteristic chart

lb/inch force, there was no metal removal rate, but after that Z'_w was linear with normal force. The 20 lb/inch is labeled F'_{th} the threshold force, because it was the force threshold that had to be overcome to cause metal removal. The slope of the Z'_w vs. F'_n line is labeled Λ_w, or the work removal parameter (WRP) and has the units: cubic inch of metal being removed per minute per pound of force, 0.014 inch³ minute, lb. The magnitude, or "steepness," of that line is important because it defines how "sharp" the system is cutting. A steep slope means fast-cutting ("free-cutting") and low forces; a shallow slope means slow-cutting ("hard-acting") and high forces. Along the bottom, a portion of the Z'_s data is shown linear with F'_n and Λ_s; the wheelwear parameter is given as 0.000555 inch³/minute, lb. When we went to higher forces, the Z'_s curve turned up sharply, and over a wide range of force, Z'_s increases rapidly.

Power is shown linear with F'_n. That is because the tangential force is just a fraction (from .025 to 0.5 or so) of the normal force (as a "coefficient of friction" would be), so since power is F_t times wheelspeed, power is linear with F'_n. Surface finish is low at low force and increases with force.

A chart like Fig. 2.3 can be used to plan a complete production grinding cycle for a workpiece. Moreover, if the linear relationships are written as equations, numbers for the cycle may be obtained. For example:

$$Z'_w = \text{WRP } (F'_n - F'_{th}) \ \frac{(\text{inch}^3)}{(\text{min,inch})} = \frac{(\text{inch}^3)}{(\text{min,lb.})} \ \frac{(\text{lb})}{(\text{inch})} \qquad (2.5)$$

$$Z'_s = \Lambda_s (F'_n - F'_{th}) \quad \text{same units} \qquad (2.6)$$

So, at any desired removal rate Z'_w the amount of force required and the corresponding wheelwear rate Z'_s could be calculated. Also, from Fig. 2.3, the power necessary to grind at Z'_w could be found as well as the surface finish on the part after grinding.

Fig. 2.3, sometimes called a "system characteristic chart," defines the relationship of all grinding factors. That one, developed using relatively short grinding times (less than 10 seconds) does not show the effect of grinding time, which could be thought of as a "third axis" into the paper. For internal grinding, where the wheel is dressed for every part (usually between rough and finish grinds), the time axis is short, only 1-part long. For larger wheels, or extended grinding times, the time axis can be substantial and the lines of Fig. 2.3 can change slope somewhat. The effect of time will be discussed later.

Most machines set an infeed rate to the slide that generates some force depending on the system springs as described earlier. In Fig. 2.3 a dotted horizontal line at $Z'_w = 0.7$ intersects the WRP line, where a vertical line to the horizontal axis gives the force 68 lb/inch. From that force, a vertical line shows the Z'_s wheelwear rate to be .03 in³/minute, inch. (So the G Ratio would be 0.7/0.03=23.3.) The same vertical line intersects the power line at 5 hp/inch and meets the surface finish line at 47 micro-inches. In that way a system characteristic chart can be used to plan a grinding cycle for a feedrate machine.

On a controlled-force machine (patented by the Heald Division of Cincinnati Milacron in 1963), a 68 lb/inch force would be set, and all other factors would be found using a vertical line from that force to its intersection with the Z'_w, Z'_s, Power and surface finish behavior lines.

Specific Power and Specific Energy

The normal force is the independent variable affecting the grinding system. In most production grinding systems, unless a dynamometer is available, the normal force is unknown. "Controlled Force" internal grinders use the normal force as input to the feedslide: that is, hydraulic force not a mechanical screw moves the feed slide. On feedrate machines power is the only easily obtainable measure of force. Using the relationship described earlier, one can calculate the tangential force:

$$\text{Power} = \frac{F_t V_s}{33000} \qquad (hp) = \left[\frac{\text{lb. ft/min}}{\text{ft.lb/min,hp}}\right]$$

or

$$F_t = \frac{33000 \text{ Power}}{V_s}$$

(2.7)

So, for example, for a 12000 fpm wheelspeed, using 20 horsepower, the tangential force is: $F_t = 33000 \ (20)/12000 = 55$ lb.

In those cases power, or tangential force, can be used as the independent variable. (If power is used and wheelspeed varies over a wide range, corrections must be made because of the effect of wheelspeed.)

The justification for using F_t for F_n is shown in Fig. 2.4. Data in graph 2.4a, from external grinding of 52100 steel with aluminum oxide-vitrified bond wheels at 12000 fpm, show a linear F'_t vs. F'_n relationship for 3 wheel grades. The coefficient here was 0.40. The Fig. 2.4b results from internally grinding hard cast iron (R_c43) workpiece when using organically bonded Silicon Carbide grinding wheels running at 5500 surface feet per minute wheelspeed. The coefficient of friction measured in these tests when using 3 different wheel grades was 0.27. Obviously, wheel grades do not change the coefficient, but the abrasive and the metal can.

So tangential force and power can be used to define another grinding parameter: Specific Power. Figure 2.5 shows the data from Fig. 2.3.

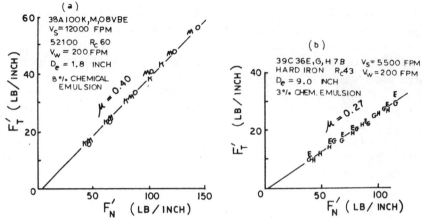

Fig. 2.4   Coefficient of grinding friction

Here, after a 1.6 hp/inch threshold power (labeled HP'$_{th}$, it corresponds to the 30 lb/inch threshold force of Fig. 2.3), power was linear with Z'$_w$ giving a slope of 5.7 hp/inch³ per minute. That means that to grind at a 1-cubic-inch-per-minute removal rate takes 5.7 horsepower (plus the threshold power). So we can write:

$$P' - P'_{th} = (\text{Specific Power})\, Z'_w \frac{(hp)}{(inch)} = \left[\frac{hp}{in^3/min}\right]\left[\frac{in^3}{min, inch}\right]$$

or

$$P' = (\text{Specific Power})\, Z'_w + P'_{th} \qquad (2.8)$$

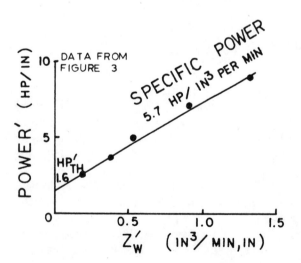

Fig. 2.5   The specific power slope

Because F'_n and F'_t are related by the coefficient of grinding friction (Fig. 2.4), specific power and WRP, the work removal parameter, are just "two sides of the same coin."

Other researchers divide each power value by its corresponding Z'_w value, obtaining a different number for each test condition: 5 numbers for 5 Z'_w values.

Those numbers (power/Z_w) are called "specific energy" or "specific grinding energy." They are the mathematical equivalent of drawing a line from the origin to each point, as Fig. 2.6a shows. Those "specific energy" numbers are usually plotted against Z'_w, as in Fig. 2.6b; thus "specific energy decreases with metal removal rate." Actually, if there were no threshold power (and normal force), all 5 values of power/Z_w would be nearly identical: 5.7 hp/inch per minute. Thus, all the right graph illustrates is that a threshold power exists for this system. "Specific energy" also implies that power decreases with higher Z'_w, and of course it does not. Specific power, as shown in Fig. 2.5, is more physically descriptive of what is happening and less confusing.

Another example where the physical significance of grinding behavior was lost by using "specific grinding energy" is shown in Fig. 2.7. Graph 2.7a shows power vs. Z_w for plunge grinding and traverse grinding with 20-inch diameter wheels at 6500 fpm. For traverse grinding I used the "cross-feed width" (axial traverse rate/work rotational speed: inch per minute/revolutions per minute) as the grinding width B. Up to about 3.5 inch³/minute inch, the plunge and traverse mode behavior connected nicely, making specific power slopes of 7.41 to 7.91 horsepower/inch³ per minute. Then the curves broke to form 3 distinct new specific power slopes: 0.023, 4.35, and 4.46 for I, K, and M grades respectively. That

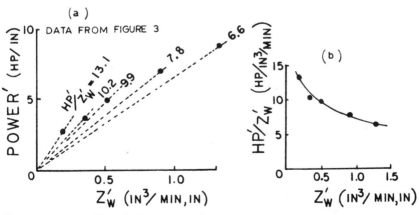

Fig. 2.6 "Specific grinding energy"

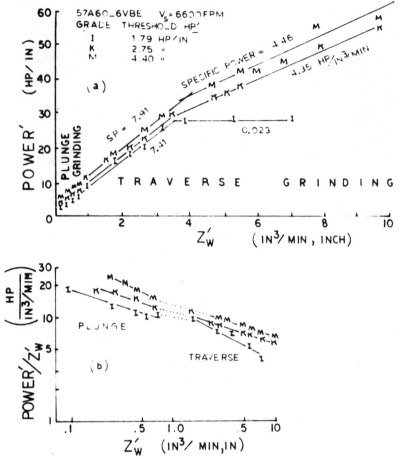

Fig. 2.7   Compare specific power and "specific energy"

happened because the wheelwear rate reached levels high enough to sharpen the wheel, making it cut with lower forces and causing the "knee" in the curve for each grade. The powers (or F_t) where the 3 curves broke were 28, 29, and 34 hp/inch (F_t' of 142, 147, and 173 lb/inch) for I, K, and M. Fig. 2.7b shows values of "power/Z_w" or "specific energy" vs. Z_w', and while that gave a clean separation and continuous curves, the physical significance of wheelwear rate sharpening and the forces where it occurred for each grade are completely lost. That picture tells us nearly nothing compared with Graph 2.7a.

H Equivalent

Still another method of presenting grinding data is shown in Fig. 2.8. Here the horizontal axis is "H equivalent," or Z'_w/V_s (giving the units "micrometer" or "microinch"). F'_n, F'_t, G Ratio, surface finish (Ra) at 60 m/s and 30 m/s (12000 and 6000 fpm) are plotted against H equivalent. That presentation is used mainly in Europe. However, it suggests that the normal force is an *effect* of the metal removal rate, not the *cause* of it. Nevertheless, all the data are significant and useful, and many charts for different materials have been generated.

I took data from Fig. 2.8 and converted the metric units, and the results are shown in Fig. 2.9, in the method of Fig. 2.3 and 2.5. The similarities of those 3 figures are obvious. That means that while the people using Fig. 2.8 type graphs do not recognize the significance of normal force as input to the grinding process, their results and production planning based on them will be valid.

The alternative methods of data analysis, "Specific Grinding Energy" and "H equivalent," are presented to acquaint the reader with them. They are widely used and supported. Neither is wrong or incorrect; they represent differences of opinion regarding the significance of normal

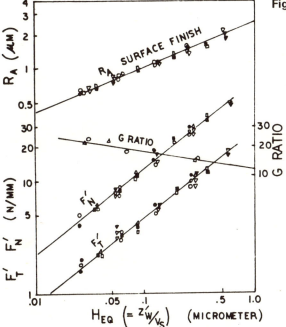

Fig. 2.8   The "Heq" chart

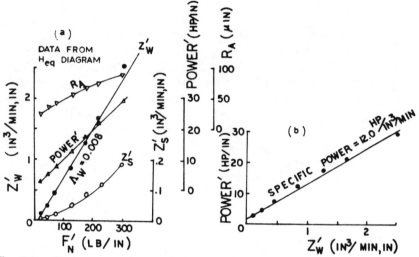

Fig. 2.9   The system characteristic chart from the "Heq" chart

force, and it was felt that this section should discuss them because they exist in the literature.

G Ratio

The G Ratio has been defined as:

$$\text{G Ratio} = \frac{\text{volume of metal removed}}{\text{volume of wheel used}} \quad \frac{(\text{inch}^3)}{(\text{inch}^3)} \quad \textbf{(2.9)}$$

But we can divide top and bottom of the right side of Eq. 2.9 by time and get:

$$\text{G Ratio} = Z'_w/Z'_s \quad \textbf{(2.10)}$$

The G Ratio has been used for a long time, and many people know its meaning and economic significance.

Fig. 2.10 shows Z'_w vs. F'_n for 3 wheel grades, where the wheels were dressed aggressively (dressing will be discussed later) and high forces were used. In Graph 2.10a the work removal parameter was between 0.018 and 0.019 inch³/minute, lb until the K and M grades began to cut faster (steeper slopes) around 100 lb/inch. That was similar to the "sharpening behavior" that was seen in Fig. 2.7, where the power vs. Z'_w data turned down at high power (force). In graph 2.10b the wheels wore the way we would expect: K, the softest wheel (has the least bond) wore the fastest at all forces, M was intermediate, and O wore the slow-

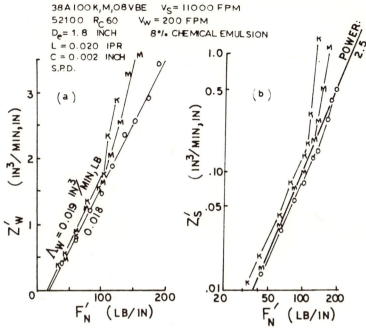

Fig. 2.10 The effect of three wheel grades on cutting and wearing

est. Notice the heavy line labeled "power = 2.5." That means Z_s' is a function of normal force to the 2.5 power: a nonlinear relationship.

Fig. 2.11 shows G Ratio vs. Z_w' for those wheels. Up to around 1.5 inch³/minute, inch the K gave the lowest G Ratios, M was best, and O intermediate. That is important: The slowest-wearing O grade did not produce the highest G Ratios. Beyond Z_w' of 1.5, the O generally became best, M intermediate, K worst. That often happens: The hardest, slowest-wearing wheel does not give the best G Ratio performance especially at low-to-moderate force levels. At high force levels or on difficult-to-grind material, the hardest grade often produces the best G Ratios.

G Ratio behavior is a complex interaction of cutting and wearing, Z_w' and Z_s' vs. F_n'.[3]

The Work Removal Parameter, WRP and Specific Power

Since the slope of Z_w' vs. F_n', the work removal parameter (WRP), is so important, we will discuss some factors influencing it.

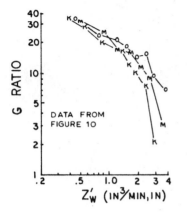

Fig. 2.11   G ratio vs. Z'_w for 3 wheel grades

Wheelspeed Effect

For most easy-to-grind materials (non-tool-steels generally), higher wheelspeed increases the WRP. Fig. 2.12 shows internal grinding data at 6000 and 12000 fpm. In Graph 2.12a raising wheelspeed increased the WRP slope from 0.0057 to 0.0129 inch³/minute, lb. Notice the threshold force also increased from 12 to 35 lb./inch. In Graph 2.12b increasing wheelspeed lowered the specific power slope from 13.9 to 11.3 hp/inch³ per minute, but it increased the threshold power from nearly zero to 4.4 hp/inch. Consequently, at any Z'_w below 1.69 inch³/minute, inch, higher power will be required at high wheelspeed. (1.69 Z'_w is where the 2 lines intersect.)

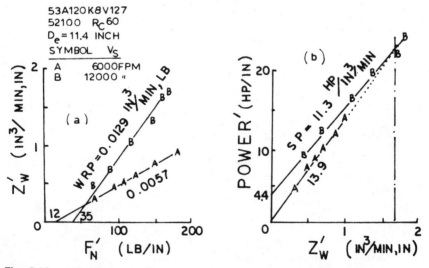

Fig. 2.12   Wheelspeed effects: internal grinding

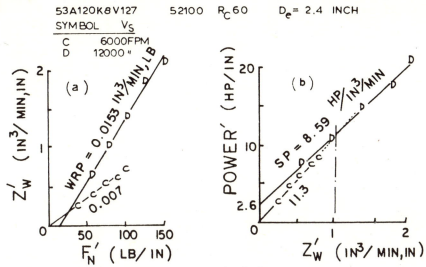

53A120K8V127   52100 R_C60 D_e = 2.4 INCH

SYMBOL	V_S
C	6000FPM
D	12000 "

Fig. 2.13   Wheelspeed effects: external grinding

Fig. 2.13 shows similar data for external grinding. In Graph 2.13a increasing wheelspeed again raised the WRP slope from 0.007 to 0.0153 inch³/minute, lb but also increased the threshold force from zero to 14 lb./inch. In graph 2.13b the specific power slope was smaller for high wheelspeed: 8.39 vs. 11.3 hp/inch³ per minute. However, as before, high speed raised threshold power from zero to 2.6 hp/inch. So, as in Fig. 2.12, at any Z'_w below 0.970 (where the two lines intersect), the high-speed system will require higher power.

Equivalent Diameter

We know in external grinding the "fit" or conformity of the wheel to the workpiece is different from that in internal grinding. One way to relate them is by the equivalent diameter:

$$D_e = \frac{(\text{Work diameter}) \, (\text{Wheel diameter})}{\text{Work diameter} \pm \text{Wheel diameter}} \, (\text{inch}) = \frac{(\text{inch}) \, (\text{inch})}{(\text{inch})} \quad (2.11)$$

Use − for internal grinding and + for external grinding.

The equivalent diameter (inch) is the diameter of a wheel that would be used for surface grinding (as in Fig. 2.2) to represent the way the wheel and work fit together.

Table 2—2 shows 5 examples:

Table 2—2

No.	Mode	DW	DS	$D_e = D_w \cdot D_s/(D_w \pm D_s)$
1	Internal	4.0	3.6	4(3.6)/(4−3.6) =14.4/0.4=36.0 inch
2	Internal	4.0	1.0	4(1)/(4−1) = 4/3 =1.33 inch
3	External	4.0	20	4(20)/(4+20) = 80/24 =3.33 inch
4	External	30.0	30	30(30)/(30+30)= 900/60 =15.0 inch
5	External	1.0	20	1(20)/(1+20) = 20/21 =0.95 inch

Obviously, De can vary widely. Example 4, approximately the system for grinding paper-mill rolls, is more like internal grinding with a De of 15.0 inch. In fact, a few years ago, I modeled paper-mill roll grinding by internal grinding in the laboratory, choosing wheel and work sizes to get an equivalent diameter of 15.0 inch.

Fig. 2.14 illustrates examples 1, 3, and 5 from Table 2—2. Obviously, for a 100-lb normal force on the 36.0-inch and 3.3-inch equivalent diameters, the contact length will be much larger for De of 36.0 inch. That

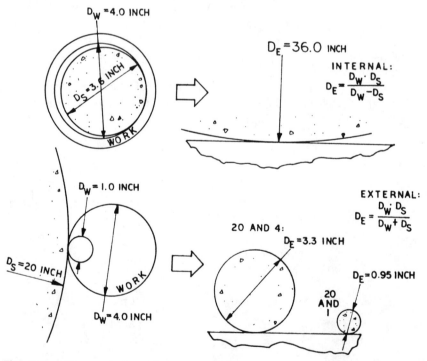

Fig. 2.14   The equivalent diameter: relating cylindrical to surface grinding

means there will be more grains in contact as De increases, and so a lower force/grain. As the force/grain decreases, the ability to start a chip and remove metal rapidly will decrease: The wheel acts duller.

Fig. 2.15 shows the high wheelspeed results from Figs. 2.12 and 2.13 to illustrate the effect of equivalent diameter. In Graph 2.15a the internal ("I" symbols) system had a threshold force of 35 lb/inch and a WRP slope of 0.0129 inch³/minute, lb. while the external system ("E" symbols) had a threshold force of 14 lb/inch and a WRP slope of 0.0153 inch³/minute, lb. Thus, the lower equivalent diameter system (2.4 inch vs. 11.4 inch) had a lower threshold force and steeper WRP slope, as expected.

In graph 2.15b the external system had a lower threshold power, 2.6 vs. 4.4 hp/inch, and a lower Specific Power, 8.59 vs. 11.3 hp/inch²/per minute. Notice there were no "crossovers" in that graph, as occurred for high and low wheelspeed.

Of course, for surface grinding as in Fig. 2.2, the equivalent diameter is the size of the wheel.

Wheel Dressing

For nonrotating diamond dressing, either with a single-point or multi-point diamond, the dressing lead and depth are two important factors, as illustrated in Fig. 2.16. The equation shows how to calculate dress lead. Obviously, just knowing the wheel was "dressed at 20 inch/minute traverse rate" is meaningless without knowing the rotary speed. In-

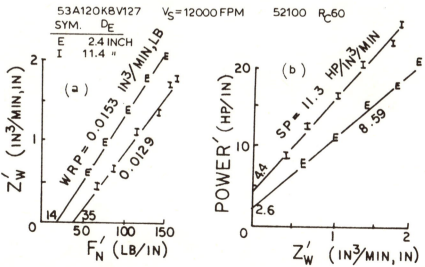

Fig. 2.15   Effect of equivalent diameter at high wheelspeed

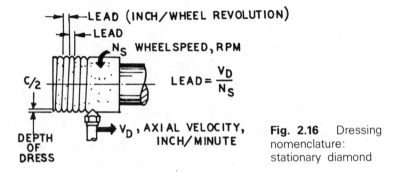

Fig. 2.16 Dressing nomenclature: stationary diamond

creasing either the lead or depth increase the "aggressiveness" of the dress and sharpens the wheel. However, it is uneconomical to increase the depth more than is necessary because the wheel is being wasted. Higher leads are more economically effective.

Fig. 2.17 shows 3 graphs illustrating the effect of changing dressing conditions from "gentle" (lead=0.004 ipr/C, diametral depth = 0.001 inch) to "harsh" (lead=0.020 ipr/C=0.002 inch). In graph 2.17a the WRP increased from 0.016 to 0.022 inch ^{3}/minute, lb. In Graph 2.17b the specific power fell from 8.0 to 6.0 hp/inch^3 per minute. In graph 2.17c, however, an economic penalty was paid, as the gentle dress gave higher G Ratios at all removal rates. Actually, the 0.004 ipr/0.001 inch dressing conditions called "gentle" here are probably a reasonable upper limit for precision grinding using a single-point diamond, although 0.020 ipr/ 0.002 inch is used, for example, on some cam grinders using 3 diamonds in a vertical row on large wheels. That condition is also common with multipoint diamonds. Rotary diamond dressing is sometimes used on high production, long-run jobs where the quantity and shape of the workpiece form merit investing in an expensive dressing system.

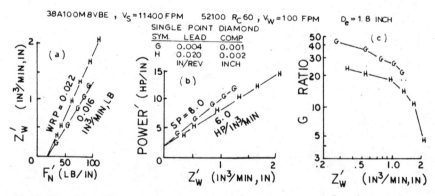

Fig. 2.17 The effect of dressing on cutting, power, and G ratio

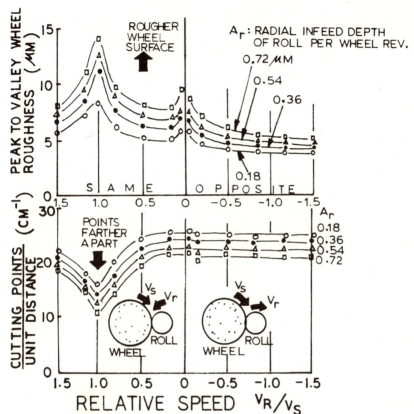

Fig. 2.18   Dressing nomenclature: parallel-axis rotary diamond

Fig. 2.18 shows a famous graph developed at the Technical University of Braunschweig, West Germany. The horizontal axis is the ratio of roll speed to wheel speed: V_r/V_s in feet/minute. There are 2 directions: To the left of zero, the wheel and roll go in the same direction at the contact zone, and to the right of zero (negative numbers) the wheel and roll go in opposite directions at contact. The +1.0 condition means wheel and roll travel at the same speed: a "crushing mode." The "zero" condition means the roll was stopped; +1.0 and zero cause the roll to wear excessively and generally are not used. The vertical axis of the upper graph is R_{ts} the roughness of the grinding wheel surface. The vertical axis of the lower graph is the number of cutting points per linear distance (in centimeters). The symbol A_r on the 4 curves is the distance the diamond roll infeeds per revolution of the wheel; it is analogous to the radial depth of dress, $c/2$ from Fig. 2.16. Obviously, increasing A_r (dressing deeper) increased the R_{ts} value and decreased the number of cutting points, so the wheel would act sharper at higher A_r values.

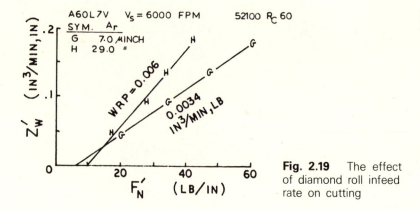

Fig. 2.19   The effect of diamond roll infeed rate on cutting

Also, switching roll *direction* to operate at +0.8 for rough grinding (to the left of zero) would produce a high R_{ts} wheel surface, which should act sharp with a steep work removal parameter slope. Then operating at −0.8 (to the right of zero) would produce a smoother wheel surface with a lower R_{ts} value, for finish grinding. Another method would be to use a high A_r value for rough-grinding and a low A_r value for finish-grinding.

Figure 2.19 shows data from the same source. Obviously, using a 0.000029 inch/revolution A_r value sharpened the wheel and produced a WRP slope of 0.006 inch³/minute, lb. A gentle infeed rate of 0.000007 inch dulled the wheel and reduced the WRP to 0.0034.

Wheel Grade

Fig. 2.20 shows the effect of wheel grade on the work removal parameter and specific power. In graph 2.20a harder grades reduced the WRP from 0.019 at K grade to 0.0145 at M and 0.0115 inch³/minute, lb for O grade. In graph 2.20b increasing grade increased the specific power slopes from 7.5 for K, to 10.0 for M and 11.5 hp/inch³ per minute for O grade. (Incidently, when similar wheels were trued with harsher dressing conditions (Fig. 2.10), the harder M and O grades were made to cut more like the K grade: WRP values were 0.018–0.019 in Fig. 2.10, but they had a wider range here.)

Coolant Effects

The subject of grinding fluids is still somewhat regarded as a mystery, probably because coolant manufacturers keep their ingredients secret and users generally do not take or have the time to evaluate new fluids. Also, the reality is that fluids—and sometimes wheels and diamonds—are usually bought by purchasing people who seek the lowest price with

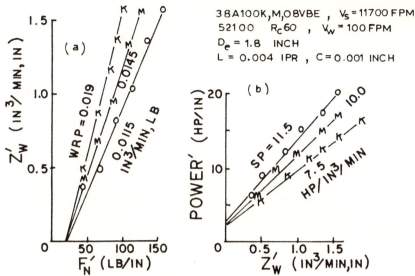

Fig. 2.20   Gentle dressing: the effect of three wheelgrades

little regard for quality or performance. That is the real world. However, fluids *can* improve grinding performance tremendously. The author is convinced that is true.

Three examples show what a higher quality/price fluid can do. Fig. 2.21 shows external grinding of hardened M2 tool steel (R_c 64) using a chemical emulsion and sulfur-chlorinated grinding oil. Graph 21a shows that both fluids provided essentially equivalent Z'_w vs. F'_n results except that because the oil system caused relatively low wheelwear rates, we

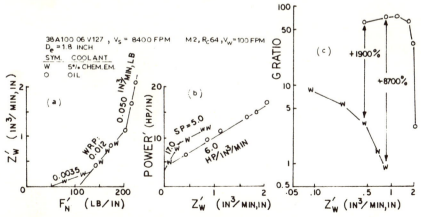

Fig. 2.21   Coolant effects: chemical emulsion vs. oil

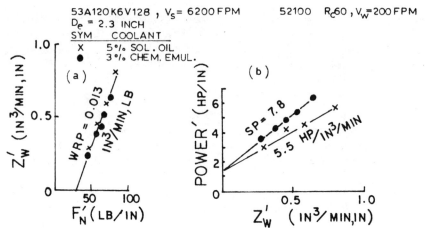

Fig. 2.22   Coolant effects: chemical emulsion vs. soluble oil (low wheelspeed)

could go to higher forces. Graph 2.21b shows oil required less power at all Z_w' values, mainly by reducing the threshold power, since specific power slopes were nearly identical. Graph 2.21c shows oil improved G Ratios tremendously: allowing higher removal rates as a bonus.

Fig. 2.22 and 2.23 show the effects of external grinding 52100 (R_c60) steel using chemical emulsion and a high-quality soluble oil at 2 wheelspeeds. Graph 2.22a (at 6200 fpm) shows the WRP slopes were essen-

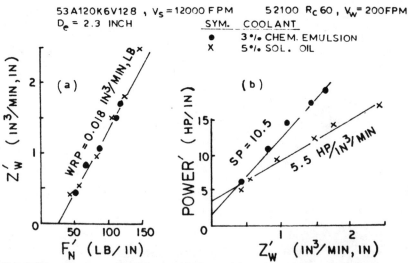

Fig. 2.23   Coolant effects: chemical emulsion vs. soluble oil (high wheelspeed)

tially equal. In graph 2.22b, however, the soluble oil reduced specific power from 7.8 to 5.5 hp/inch³ per minute: 29.5%. Graph 2.23a (for 12000 fpm) shows equivalent WRP slopes. Graph 2.23b again shows a sharp reduction in power requirements (47.6% here) using the soluble oil.

So, in 3 cases, there were substantial reductions in power requirements using higher-quality fluids. Users fighting surface integrity requirements (metallurgical damage or residual stresses) should be aware that *anything* that reduces forces and power in grinding will help them. Higher-quality, and probably more expensive fluids can reduce forces and power.

The Work Removal Parameter Equation

As part of the author's doctoral thesis[5] an equation was developed predicting the work removal parameter, WRP, using all the factors for a particular grinding system. At least 1 large bearing manufacturer uses that equation sucessfully (and others from the thesis) to plan its production grinding cycles. The equation for easy-to-grind steels (nontool steels) using single-point diamonds is:

$$ \text{WRP} = \left[\frac{0.021 \text{ inch}^{819/304}}{\text{lb,ft}} \right] \left[\frac{\left[\frac{V_w}{V_s} \right]^{3/19} \left[1 + \frac{2}{3} \frac{C}{L} \right] L^{11/19} V_s}{D_e^{43/304} (\text{VOL})^{0.47} d^{5/38} R_c^{27/19}} \right] $$

Where WRP = inch³/minute, lb
V_w, V_s = ft/minute
L = inch per wheel revolution
D_e = inch
d = grain size in the wheel (inch)
Vol = 1.33H + 2.2S-8, volume factor for wheel
C = diametral depth of dress (inch)
R_e = Rockwell Hardness

Where wheel hardness is H, I, J, K, L, M, etc., the

Value of H is 0, 1, 2, 3, 4, 5, etc.
S is wheel structure number of wheel, 4, 5, 6, etc.

For example calculate the expected WRP for internal grinding a $R_c 60$, 52100 steel ring of diameter 3.75 inch, with a grinding wheel of 3.25-inch diameter, running at 12500 fpm. Workspeed is 300 fpm. Wheel is 80K5V. Dress lead will be 0.004 ipr and dressing compensation 0.001 inch:

$$\text{WRP} = \frac{(0.021) \left[\dfrac{300}{12500}\right]^{3/19} \left[1 + \dfrac{2}{3}\dfrac{(.001)}{(.004)}\right] (.004)^{11/19} (12500)}{\left[\dfrac{3.75\,(3.25)}{3.75\,-3.25}\right]^{43/304} \left[1.33(3) +2.2(5)-8\right]^{0.47}(.010)^{5/38}(60)^{27/19}} \quad (2.12)$$

WRP = 0.00968 in³/minute, lb.

Since $F_t/F_n = \mu$, the coefficient of grinding friction, the specific power is:

$$\text{SP} = \frac{\mu V_s}{33000\,(\text{WRP})} \quad\quad (2.13)$$

Where μ = coefficient of grinding friction

$$V_s = \text{fpm}$$
$$\text{WRP} = \text{inch}^3/\text{minute, lb.}$$

So, estimating μ as 0.5, the specific power for the above WRP is:

$$\text{SP} = \frac{0.5\,(12500)}{33000\,(0.00968)}$$

SP = 19.6 hp/inch³ per minute.

With μ estimated as 0.30, specific power is:

$$\text{SP} = \frac{0.3\,(12500)}{33000\,(.00968)}$$

SP = 11.7 hp/inch³ per minute

The values 0.5 and 0.3 are representative values for a chemical emulsion and soluble oil, as calculated from the data of Fig. 2.22 and 2.23, using Eq. 2.13. That convincingly shows the effect of a "lubricating fluid" as opposed to a "cooling fluid." (The μ for Fig. 2.20, using oil was 0.28).

Time Behavior

It has been stated many times that "grinding wheels dull with pro-longed usage." They will (CBN wheels do not generally dull; they are in a separate category) if the force is low enough to ensure wear by attri-tion. If the force is high enough to cause grain-fracture wear, the wheels may grind steadily if fracture sharpening balances grain dulling. In fact, if the forces are very high, the wheel will cut faster with time, as sharp-ening overwhelms flat-growth. The intermediate condition, "self-sharpening" is what many users strive for.

Fig. 2.24 gives 3 graphs for external cylindrical grinding 4340 (R$_c$43) steel using 3 grades of 60-grit wheels over extended grinding times (gen-

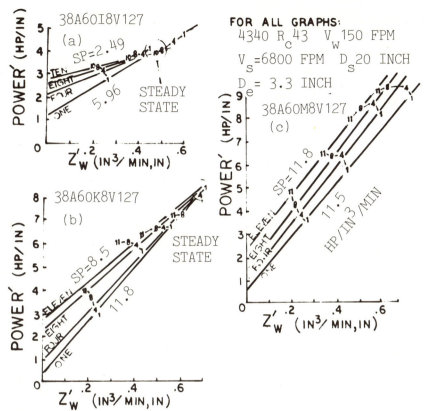

Fig. 2.24   Time behavior: external grinding with 3 wheel grades

erally removing 5.5 inch³/inch of metal). The I grade is shown in graph 2.24a. At the lowest Z_w' the numbers 1, 4, 8, and 10 (which are the number of grinds made, each removing about 0.5 inch³/inch) show power increased from the beginning (grind 1) to the end (grind 10) (dressing was before the first grind). For the next 2 Z_w' rates, grinds 10 had lower power than grind 1: The wheels *did not* dull with time. Connecting grinds 1, 4, 8, and 10 at each condition produced 4 lines, which intersected at around 0.5 inch³/minute, inch and 4 hp/inch. Theoretically, grinding at that condition would produce no dulling *or* sharpening, just steady grinding.

Graph 2.24b shows the K grade, where 11 grinds were made at four Z_w' values. Connecting the like-number grinds (1, 4, 8, and 11) produced a "fan-shaped" wedge, intersecting at about 0.7 inch³/minute, inch and 8.5 hp/inch. So the harder K grade required a much different steady condition: 8.5 hp/inch vs. 4 hp/inch for the I.

Graph 2.24c shows the hardest wheel, an M grade. Connecting grinds 1, 4, 8, and 11 here produced 4 nearly parallel lines: There was no intersection, and that wheel will *not* self-sharpen within our test range; it will *always* dull, causing rising power on infeed rate machines.

Other data confirm that behavior. Generally, with the use of coarse grit (46, 60), wheels just about guarantee dulling at all Z'_w values. Hard grades (L, M, and harder) will, too. Finer grits (100–120, etc.) and softer grades (I, K, etc.) will help a user reach a steady-state grinding condition. With proper wheel selection and a devoted effort to study their grinding cycle, a user could establish steady-state conditions on extended use (without dressing) operations.

Belt Grinding

The discussions have emphasized wheel grinding so far, mainly because belt grinding results generally do not include normal force. Recently, however, researchers have provided sufficient data to obtain the work removal parameter and specific power.[7] Figure 2.25 shows belt grinding data for 3 materials. In Graph 2.25a the WRP slopes were 0.0478 inch³/minute, lb for cast iron, 0.0312 for 1018 steel and 0.0108 for 304 stainless steel. The specific power slopes in Graph 2.25b were 2.20 hp/inch³ per minute for cast iron, 3.30 for 1018 steel, and 6.59 for 304 stainless. The interesting thing is that those results conclusively show belt grinding behaves like wheel grinding. Moreover, making the backup roll smaller or larger or softer or harder (rubber vs. steel, say) causes the same changes in belt grinding that the equivalent diameter does for wheel grinding.

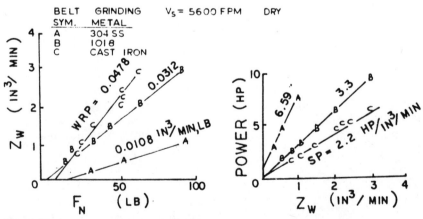

Fig. 2.25   Belt grinding 3 materials

Abrasive Cutoff

Just as abrasive belt grinding is a rapid metal removal process, abrasive cutoff is used to quickly chop steel into specific lengths. Fig. 2.26 shows the results of research done at Carnegie Mellon University for the Grinding Wheel Institute.[8] In Graphs 2.26a, metal removal rate vs. normal force, and 2.26b, power vs. metal removal rate, wheel A36R6B cut best: highest WRP slope and lowest specific power slope. The other 4 wheels were virtually identical in those graphs. However, in Graph 2.26c, G Ratio vs. metal removal rate, wheel performance was clearly defined. Since a customer could use any removal rate, the uppermost wheel guarantees the highest G Ratios, so the A24R6 wheel was best. Notice that the fastest-cutting/lowest power wheel, A36R6B, was exactly intermediate in this graph. The wheels were aligned clearly: 24R, 24P,

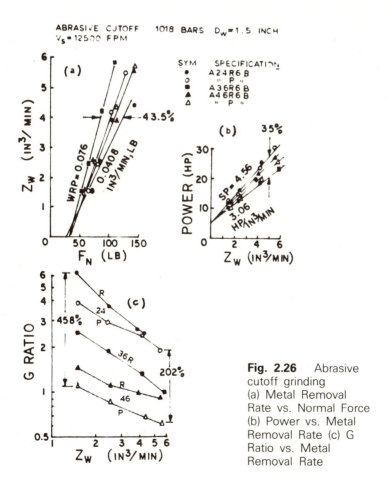

Fig. 2.26 Abrasive cutoff grinding (a) Metal Removal Rate vs. Normal Force (b) Power vs. Metal Removal Rate (c) G Ratio vs. Metal Removal Rate

36R, 46R, and 46P; largest grit-hardest grade best, harder grade always better, and finest grit-softest grade worst.

Sparkout

At the end of most precision grinding cycles is a "sparkout," where the cross-slide motion stops and the normal force between the wheel and work is allowed to decrease by letting the system "grind itself free." During sparkout the normal and tangential forces decrease, metal is removed at a slower and slower rate, and surface finish and geometry continuously improve. The sparkout process has been shown to follow a mathematical relationship[9]:

$$F_n' = F_o'e^{-t/\tau} \tag{2.14}$$

where $\tau = \pi D_w B/K \cdot WRP$ (minute) $= \dfrac{\text{(inch) (inch)}}{\text{(lb/inch)(in}^3\text{/min,lb)}}$ (2.15)

and K is the effective spring stiffness of the grinding system; e is the number 2.718 and t is the time.

In 3 time constants ($t = 3\tau$), the grinding system will reach nearly steady state, and further time is just wasted. For example, for a 4-inch diameter part, 0.5 inch wide with a Work Removal Parameter of 0.010 inch³/minute, lb. and a spring stiffness of 50,000 lb./inch, τ is:

$$\tau = \frac{\pi(4) \quad (0.5)}{(0.010) \ (50,000)} = 0.0126 \text{ minute} = 0.75 \text{ seconds}$$

If the grinding force just before sparkout began was 50 lb/inch (F_o'), then the normal force at any sparkout time is:

$$F_n' = 50e^{-t/.75}$$

or:

Table 2—3

t (secs)	$\dfrac{t}{\tau}$ t/0.75	$e^{t/\tau}$	$e^{-t/\tau}$	F'_n
0	0	1.00	1.00	50.0
0.75	1.0	2.718	0.368	18.4
1.5	2.0	7.387	0.135	6.8
2.25	3.0	20.08	0.0498	2.5
3.0	4.0	54.58	0.018	0.9

Table 2-3 shows that in 3.0 time constants (2.55 seconds of sparkout) the 50 lb./inch initial force has decreased to 2.5 lb./inch, certainly very close to zero, the steady state. The powerful effect of the WRP and system stiffness on the time constant can be seen from cutting both values in half:

$$\tau = \frac{4}{0.005} \frac{(0.5)}{(25000)} = 0.0502 \text{ minutes} = 3.0 \text{ seconds}$$

To reach 2.5 lb/inch grinding force with this system would require 9.0 seconds (3 time constants).

So large diameter (D_w) or wide (B) workpieces, stiff systems (high K value) and sharp-acting (high WRP values) wheel systems are important or sparkout times will become excessively long.

The time constant (Eq. 2.15) can be rewritten using Eq. (2.13) solved for WRP:

$$WRP = \frac{\mu V_s}{33000SP} \frac{(\text{inch}^3)}{(\text{min,lb})} = \frac{(\text{ft})}{(\text{m/n})} \frac{(\text{min,hp})}{(\text{ft,lb})} \frac{(\text{inch}^3)}{(\text{hp,min})} \text{ into } \tau:$$

$$\tau = \frac{\pi D_w B \, 33000 \, SP}{\mu V_s K} \qquad (2.16)$$

Eq. (2.16) could be used on a production system where specific power had been measured using an estimate of the coefficient of friction.

Difficult-to-Grind Steels

The discussion so far has been about easy-to-grind steels, that is, generally, nontool steels. And even among tool steels there are degrees of difficulty for grinding and machining. The high-speed steels, M and T grades, are generally the most difficult to machine and grind; for example: T15, M4, M42. Early work on high-speed steels showed the presence of carbide particles whose hardness (Knoop 2300–2700) approximated that of the abrasive grits (1900–2900). The alloying elements vanadium and chromium, when present in percentages of 4.0 and above, create a large quantity of those carbide particles, making those materials very difficult to grind with non-CBN wheels.[1]

Fig. 2.27, 2.28, and 2.29 show internal (large De values) and external (small De values) grinding results on hardened M4, T15, and M50 using oil. As De was increased (larger wheelwork contact length), the threshold forces necessary to begin cutting increased sharply for all 3 materials. Fig. 2.30 shows the threshold force dependency on Equivalent Diameter for those metals and 52100 (ball bearing steel). Obviously for high-speed steels and high wheelspeed, high threshold forces will exist, much higher than for 52100 or other easy-to-grind materials.

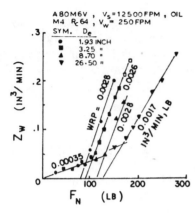

Fig. 2.27   Equivalent diameter effects grinding M4 steel. Metal removal rate vs. normal force for several values of equiv. diam.

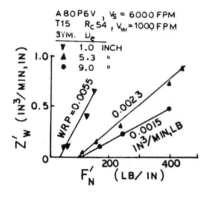

Fig. 2.28   Equivalent diameter effects grinding T15 steel. Normalized removal rate vs. force intensity

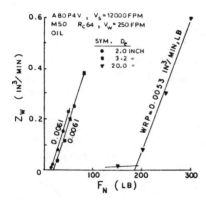

Fig. 2.29   Equivalent diameter effects grinding M50 steel. Metal removal rate vs. normal force for 3 values of equivalent diameter

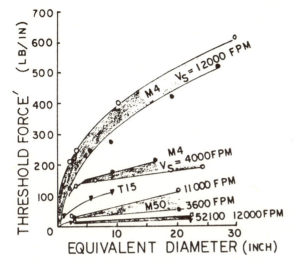

Fig. 2.30 Threshold force intensity as a function of equivalent diameter. Threshold force intensity vs. equivalent diameter for T-15, M-4, M50 at several speeds.

Fig. 2.31 shows that even non-high-speed steels can often be "difficult-to-grind." In Graph 2.31a the 2 metals, 52100 (R_c60) and 440 stainless steels (R_c 57), had nearly equivalent WRP slopes: 0.0092 and 0.0096 inch³/minute, lb, although the threshold force difference (20 vs. 36 lb./inch) meant that 440 SS would always require higher forces. Graph 2.31b illustrates that the higher threshold power meant 440 SS would

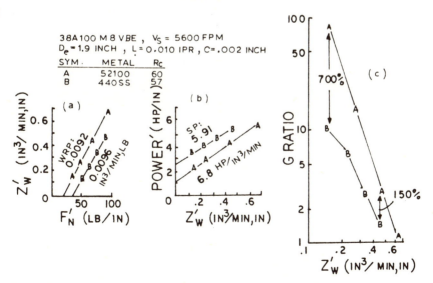

Fig. 2.31   Comparing 440 stainless with 52100 steel: low wheelspeed

use higher power even though its specific power slope was smaller: 5.91 hp/inch³ per minute vs. 6.8. (Those yield μ values of 0.334 for 440 SS and 0.368 for 52100.) In Graph 2.31c G Ratio vs. Z'_w shows that 440 SS caused much higher wheelwear rates and lower G Ratios: 52100 had from 150% to 700% higher G Ratios than 440 Stainless did, even though the first 2 graphs showed the 2 metals ground similarly. Since 440 stainless looked interesting, we increased the wheelspeed to 10600 fpm, dressed less severely (lead of 0.004 ipr, C=0.001 inch on diameter compared with 0.010 ipr/0.002 inch for Fig. 2.31), and changed to a K grade wheel. Fig. 2.32 shows those results (with the Fig. 2.31 data shown with an "a" symbol) for the rust inhibitor fluid for Fig. 2.31 and a chemical emulsion fluid. Graph 2.32c shows we increased G Ratios substantially over the low wheelspeed results, but now, at high wheelspeed, we need more power to grind, as seen in Graph 2.32b. Graph 2.32a shows that the chemical emulsion fluid substantially increased the WRP slope and required a lower threshold force, too. With the rust inhibitor fluid, the harder grade/lower wheelspeed/aggressive dress WRP (0.0096) was essentially equal to the softer-grade/higher wheelspeed/gentler dress WRP: 0.0105 inch³/minute, lb.

So, while 440 stainless steel (R_c57) probably is not regarded as "difficult-to-grind" as high-speed steels, it was more difficult than 52100 (R_c60). And we could substantially improve the grinding behavior of 440 SS by using higher wheelspeed and a better grinding fluid.

The Wheelwear Parameter

As discussed previously, wheelwear is a sum of attritional (grain-flat wear) and grain-fracture wear, and since the latter type increases with force, the wheelwear rate vs. force relationship is nonlinear. For easy-to-

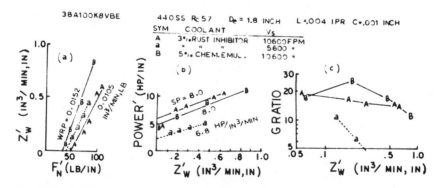

Fig. 2.32   440 stainless with 2 wheelspeeds and 2 coolants

grind steels, the equation for Λ_s the wheelwear parameter is:

$$\Lambda_s = \frac{\left(A\frac{in^2}{lb^2ft}\right) L^2 \left(1+\frac{c}{L}\right) F'_n V_s}{De^{1\cdot2/\,vol}\,(vol)^{0\cdot85}} \left(\frac{in^3}{min.lb}\right) = \left(\frac{in^2}{lb^2ft}\right)\left(in^2\right)\left(\frac{lb}{in}\right)\left(\frac{ft}{min}\right)$$

(2.17)

For the example done earlier if the A value is 0.0002:

$$\Lambda_s = \frac{0.0002\,(.004)^2 \left[1 + \dfrac{.001}{.004}\right] F'_n(12500)}{24.375^{(1\cdot2/\,6\cdot99)} * 6.99^{(\cdot85)}}$$

$$\Lambda_s = 5.55 \times 10^{-6} F'_n$$

For $F'_n = 100$ lb/inch: $\Lambda_s = 0.000555$ in.^{3}/min, lb or since $Z'_s = \Lambda_s F'_n$:

$$Z'_s = 0.0555 \text{ in}^3/\text{min, inch at } 100 \text{ lb/inch}$$

With Eq. (2.6) the radial wheelwear rate for any size wheel could then be calculated with that Z'_s value.

As seen in previous sections, the work material and coolant can change G Ratios and wheelwear rates substantially, and no corresponding equation exists for difficult-to-grind steels.

Surface Finish

The surface finish (Ra) of a workpiece is usually specified to a certain limit, especially in precision grinding. As seen earlier, surface finish increases with normal force, so, to produce smooth finishes, the final force must be low. Wheel dressing (lead and compensation depth) was seen to strongly influence the wheel sharpness or work removal parameter (Eq. 2.12) and the wheelwear parameter Eq. (2.17). They also affect surface finish. Fig. 2.33[1] shows that as the compensation/lead ratio was decreased, finish improved at L = 0.004 ipr (Graph 2.33a) and 0.001 ipr (Graph 2.33b) at all force levels. For any dressing condition, finish increased with higher forces.

Another way of showing that is to use the average chip thickness existing during a grind[10], given below:

$$Tave = \left[\frac{14.8\text{ ft}^{19/27}}{inch^{35/27}}\right]\left[\frac{V_s}{V_w}\right]^{3/27} \frac{(dL)^{16/27}\;(1+\frac{c}{L})}{D_e^{8/27}} \left[\frac{Z'_w}{V_s}\right]^{19/27}$$

(2.18)

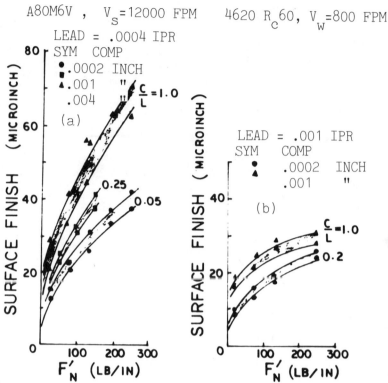

A80M6V , V_s=12000 FPM   4620 R_c60, V_w=800 FPM

Fig. 2.33   The effect of dressing parameters and normal force on surface finish

Using the previous example, where:

$$Z'_w = WRP\ F'_n = 0.0096\ (100)$$

$$Z'_w = 0.96\ in^3/min, inch$$

$$Tave = 14.8 \left[\frac{12500}{300}\right]^{3/27} \frac{(0.010 \cdot 0.004)^{16/27} \left[1 + \frac{.001}{.004}\right] (0.96)^{19/27}}{(24.375)^{8/27} \qquad (12500)^{19/27}}$$

Tave = 34.1 × 10⁻⁶ inch.

Fig. 2.34 shows how Tave was able to unify into a single relationship grinding data obtained over a wide range of operating conditions. That means to obtain some surface finish, some value of Tave (from Eq. 2.18) must be developed.

In the two regions of Fig. 2.33, surface finish is a function of Tave, as shown in the following Table:

Table 2—4

Tave Range	Ra Equation	
1–10 microinch	$Ra = 6\ Tave^{0.30}$	**(2.19a)**
10–100 microinch	$Ra = 2.2\ Tave^{0.72}$	**(2.19b)**

For the example above, Tave = 34 microinch, then the expected surface finish would be from Equ. (2.19b):

$$Ra = 2.2\ Tave^{0.72} = 2.2\ (34)^{0.72}$$
$$Ra = 28.0\ \text{microinch}$$

Obviously, to achieve an 8-microinch finish requires solving equations 2.19a backward:

$$Ra = 8 = 6\ Tave^{0.30}$$

$$1.33 = Tave^{0.30}$$

$$Tave = 1.33^{1/0.30} = (1.33)^{3.333}$$

$$Tave = 2.59\ \text{microinch}$$

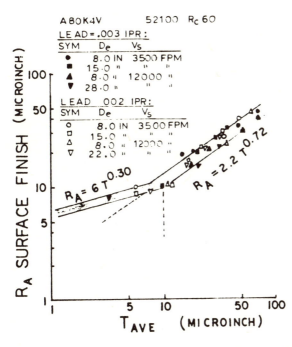

Fig. 2.34 The "average chip thickness" unifies finish data

So, to generate an 8-microinch finish, the final grind must be done using a Tave of 2.59 microinch. If one uses that value, the corresponding value of Z'_w for our system may be found as follows:

$$\text{Tave} = 2.59 \times 10^{-6} = \frac{14.8\ (1.513)\ (0.00247)\ (1.25)\ (Z'_w)^{0.704}}{(2.574)\ (765.99)}$$

$$(Z'_w)^{0.704} = 0.07386$$

$$Z'_w = 0.0247\ \text{in}^3/\text{minute, inch}$$

To grind a 4-inch diameter part at this rate requires a radial grinding rate $\bar{v}_w$, as shown below

$$Z'_w = \pi D_w\ \bar{v}_w\ 60 = 0.0247$$

$$\bar{v}_w = \frac{0.0247}{\pi\ (4)\ (60)}$$

$$\bar{v}_w = 0.000033\ \text{inch/second infeed rate}$$

That would generate an 8-microinch surface finish with no sparkout.

Adding sparkout would improve the finish, so a faster infeed rate could be used, followed by a sparkout, but that procedure supplies a starting condition for a finish-grind operation.

Surface Integrity

The quality of the surface on the finished part has recently become more and more significant. Metallurgical damage and tensile stresses in the skin of a part can cause premature failure of a whole assembly. And if that assembly is critical to a bearing life or an airplane wing or landing gear, the situation can become life-threatening. The concept of surface integrity was invented at Metcut Research Associates in Cincinnati, Ohio, which has studied causes and effects in milling, turning, grinding, and many other metal-removal processes.

$$\text{Power} = \frac{\mu F_n V_s}{33000}$$

or, writing $F = Z_w/\text{WRP}$ into above:

$$\text{Power} = \frac{\mu Z_w V_s}{33000\ \text{WRP}} \qquad (HP) = \frac{(\text{in}^3)(\text{ft}(\text{min,hp})(\text{min, lb})}{(\text{min})(\text{min})(\text{ft lb})(\text{in}^3)} \quad (2.20)$$

If we ignore threshold forces and assume WRP is proportional to V_s, then the power and consequently thermal damage will be independent of V_s. But if WRP does not increase proportionally with V_s then the power and consequently thermal damage will increase with increasing V_s.

Improving the grinding fluid to a lubricating-type straight oil or a good soluble oil (10% concentration) was shown to reduce power in Fig. 2.21, 2.23, and 2.32. So a better fluid will certainly help.

The 3 graphs of Fig. 2.35 show the effect of increasing workspeed on the residual stress in the workpiece. Plotted is the tensile (+) or compressive (−) stress in the part vs. depth below the surface. The upper graph 2.35a shows a high tensile stress at 160 fpm. Raising workspeed to 480 fpm (graph 2.35b) and 1040 fpm (graph 2.35c) produced beneficial compressive stresses.

When 52100 steel is immersed in hot (150 deg. F) hydrochloric acid for 10 minutes, hydrogen is absorbed at the surface embrittling it. If tensile stresses exist in the part, the brittle material can not elastically withstand the stress and cracks develop. To evaluate the effect of workspeed and the work removal parameter, a series of parts were ground and etched as described. At various workspeeds parts were ground at successively

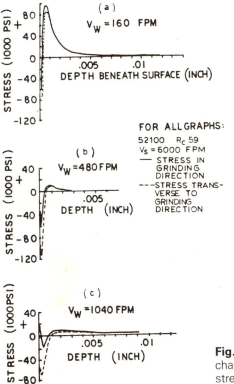

FOR ALL GRAPHS:
52100 R_c 59
V_s = 6000 FPM
— STRESS IN GRINDING DIRECTION
--- STRESS TRANSVERSE TO GRINDING DIRECTION

Fig. 2.35   High workspeed changes residual tensile stresses to compressive

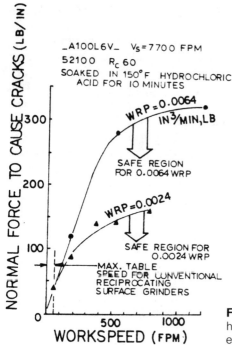

Fig. 2.36 Sharp wheels and high workspeed reduce the etch-cracking of bearing steel

higher normal forces and etched after each grind. For example, at 400 fpm workspeed, cracks began at 120 lb/inch and continuously got worse at higher forces. Finally, Fig. 2.36 was generated, where the force to cause thermal cracks is shown versus workspeed for 2 WRP values: 0.0064 and 0.0024 inch /minute, lb.

1. Thus, for a sharp (WRP=0.0064) wheel, if workspeed is 200 fpm, only F_n' less than 120 lb/inch may be used safely because higher forces are above the line and so will cause cracking. So Z_w'=(0.0064)(120)=0.768 in³/minute, inch.
2. If V_w can be increased to 500 fpm. F_n' up to 265 lb/inch could be used (while remaining beneath the curve at 500 fpm). So a maximum Z_w'=(0.0064)(265)=1.696 inch³/minute, inch.

For a dull wheel (WRP=0.0024),

1. At 200 fpm, about 90 lb/inch is safe (beneath the lower curve), so a maximum Z_w' = (0.0024)(90)=0.216 inch³/minute, inch.

WRP = equivalent dia
(wheel speed)
cutting stiffness

2. At 500 fpm, the maximum safe force is about 140 lb./inch, so the maximum $Z'_w = (0.0024)(140) = 0.336$ inch³/minute, inch.

The obvious advantage in productivity by grinding *safely* at high WRP *and* high workspeed is obvious.

Part of the problem is wheel dulling, as demonstrated in Fig. 2.24, where power (and force) rises with time at constant Z'_w. That means a higher specific power or a lower WRP from Eq. (2.23). For example, if we use Fig. 2.36, operating at 500 fpm workspeed, a 200 lb/inch F'_n is *below* the upper WRP=0.0064 curve, and so it is safe. If through use, the wheel dulls so WRP falls to 0.0024 (the lower curve), the same 200 lb/inch force is now *above* that curve and, therefore, it is *not* safe and will cause thermal damage (tensile stresses).

Another series of tests was performed for General Electric, and the workpieces were analyzed by Metcut. Here the material was Rene 80 (a high-nickel-content metal), and we used 11000 fpm wheelspeed and sulfur-chlorinated oil. The four most significant tests are shown in Fig. 2.37, residual stress vs. depth into the workpiece surface. Tests 2 and 3 produced tensile stresses of 200,000 and 60,000 psi, but tests 7 and 6 produced compressive stresses of 60,000 and 70,000 psi. Fig. 2.38 shows how we did it. Here the maximum stresses tensile (+) and compressive

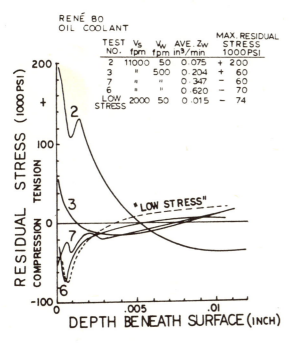

RENÉ 80
OIL COOLANT

TEST NO.	V_s fpm	V_w fpm	AVE. Z_w in³/min	MAX. RESIDUAL STRESS 1000 PSI
2	11000	50	0.075	+ 200
3	"	500	0.204	+ 60
7	"	"	0.347	− 60
6	"	"	0.620	− 70
LOW STRESS	2000	50	0.015	− 74

Fig. 2.37 Four test grinds showing high tensile and compressive stresses

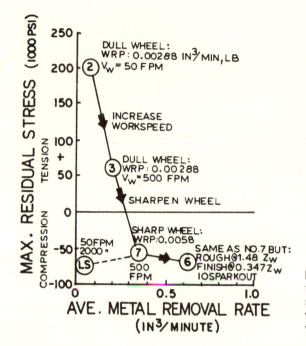

Fig. 2.38 Changing tensile to compressive stress with sharp wheels and high wheelspeed

(-) are shown as a function of the average metal removal rate during the test. Condition 2 used a wheel dulled by slow dressing (WRP of 0.00288 inch³/minute, lb) and 50 fpm workspeed. We kept the dull wheel but increased workspeed to 500 fpm and reduced the tensile stress to 60,000 psi at condition 3.

Then we dressed faster, increasing WRP to 0.0058 and kept the high workspeed. We still ground faster and produced a 60,000 psi compressive stress at condition 7. Condition 6 was a still faster metal removal rate with a sharp wheel and 500 fpm workspeed plus a 10-second sparkout (which I think cold-worked the work surface much as a shot-blasting operation would), producing a 70,000 psi compressive stress. Metcut's "low-stress" grinding recommendation (2000 fpm wheelspeed, 50 fpm workspeed, and a very low metal removal rate) is shown as the "LS" point. It produced a 74000 psi compressive stress, slightly lower than condition 6 but at a big decrease in productivity.

So, in order to improve Surface Integrity, these factors are significant:

1. Reduce forces and power by using sharp wheels that have a large work removal parameter and small specific power. That can be done, as we have seen, by using soft wheel grades, aggressive dressing, small De values, and small grit sizes.

2. High workspeed reduces wheelwork contact time, lowering the amount of heat that can flow into the work to cause damage.

3. Oil, or at least a high concentration (10%) of a good-quality soluble oil. It provides "lubrication rather than "cooling" to the grind and, as we have seen, will lower specific power slopes, reducing power.

4. High wheelspeed may increase the work removal parameter, lowering normal forces, but, as Fig. 2.12 and 2.13 showed, it may also increase the required power by raising the threshold power values. To be safer, one should avoid really high (greater than 10,000 fpm) wheelspeeds.

Cubic Boron Nitride

A complete chapter will be devoted to "Superabrasives"Diamond and CBN but a few graphs here will show that the system parameters look just like those illustrated in this chapter. CBN, a manufactured (and very expensive) abrasive, is second to diamond in hardness. Unlike diamond, however, it grinds (and cuts) steels very well. CBN is very durable and will grind tremendous quantities of steel with only very slight geometry or surface finish changes when the wheel is correctly matched to the job.

Fig. 2.39 shows four graphs linking the grinding behavior of CBN to what we have seen already. In graph 2.39a the work removal parameter was influenced by wheelwork conformity and wheelspeed. With internal grinding (equivalent diameter of 7.46 and 7.26 inch), increasing wheelspeed raised the WRP from 0.0091 to 0.0127 inch³/minute, lb. Then at 11700 fpm, reducing the equivalent diameter from 7.46 to 1.7 inch (internal to external grinding) reduced the threshold force even further and increased the WRP slope to 0.0163 inch³/minute, lb. (this was previously seen in Fig. 2.12, 2.13, and 2.15).

Fig. 2.39b shows that with internal grinding, raising wheelspeed increased the specific power slope from 7.93 to 9.38 hp/inch³ per minute. At high speed reducing conformity (internal to external grinding mode) decreased the SP slope to 7.35 hp/inch³/minute (previously illustrated in Fig. 15).

Fig. 2.39c shows the high-speed systems were relatively unaffected by conformity and gave much higher G Ratios than did low wheelspeed. Notice the high G Ratios here.

Using a modified version of Eq. 2.18 chip thickness values can be calculated for those tests, and figure 2.39d illustrates the results, just like Fig. 2.34. Notice the 2 regions here, as in Fig. 2.34, and the power of the

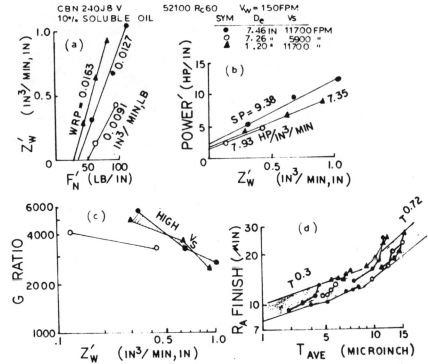

Fig. 2.39   Internal and external grinding with cubic boron nitride (CBN)

slopes were the same: 0.3 in the low-chip-thickness region and 0.72 in the high-chip-thickness range.

Those data show that the parametric behavior of CBN systems is identical to that of conventional systems, except that the G Ratios are much higher and the wheels last much longer.

REFERENCES

1.   R. S. Hahn and R. P. Lindsay, "Principles of Grinding" 5 Part Series in *Machinery Magazine*, new York, N.Y., 1971.

2.   S. Malkin, N. Joseph: "Minimum Energy in Abrasive Processes," *Wear Magazine*, 32 (1975) 15–23, Elsevier Sequoia, S.A., Lausanne, Switzerland, 1975.

3.   R. P. Lindsay, "The Effect of Wheelwear Rate on the Grinding Performance of Three Wheel Grades," *Annals of the C.I.R.P.* Vol. 32/1/1983, pp. 247–49, C.I.R.P. Pris, France, 1983.

4.   R. Snoeys, J. Peters, "The Significance of Chip Thickness in Grinding" *Annals of the C.I.R.P.*, Vol. 23, 1974, pp. 227–37, C.I.R.P., 1974

5.   R. P. Lindsay, "On the Metal Removal and Wheel Removal Parameters Surface Finish, Geometry and Thermal Damage in Precision Grinding," *Ph.D. Thesis*, Worcester Polytechnic Institute, Worcester, Mass. 1971.

6.   R. P. Lindsay, "The Effect of Contact Time on Forces, Power and Metal Removal Rate in Precision Grinding", *International Grinding Conference*, Lake Geneva, Wis., Society of Manufacturing Engineers, Dearborn, Mich., 1984.

7.   E. J. Duwell, G.R. Abrahamson and R. J. Cosmano, "Dynamics of Grinding with Coated Abrasive Belts", *International Grinding Conference*, Lake Geneva, Wisc., SME, 1984.

8.   M. C. Shaw, "Grinding Fundamentals", *Carnegie-Mellon University Grinding Wheel Institute*, Report #1, 1964.

9.   R. P. Lindsay, "Sparkout Behavior in Precision Grinding," *SME Paper No. 72 205*, Society of Manufacturing Engineers, Dearborn, Mich.

10.   R. P. Lindsay, "On the Surface Finish Metal Removal Relationship in Precision Grinding", *ASME Paper No. 72WA/Prod-13*, American Society of Mechanical Engineers, New York, N.Y., 1972.

11.   R. S. Hahn and R. P. Lindsay: "The Production of Fine Surface Finishes While Maintaining Good Surface Integrity by Grinding" *International Grinding Conference*, Carnegie Mellon University, Pittsburg, Penn., 1972.

CHAPTER 3

Types of Grinding Wheels

William Ault

Introduction

The original grinding wheel fashioned of emery and clay and turned and fired by Frank B. Norton in the late 1800s can be characterized as consisting of abrasive, bond, and air. That simple model, while complicated by the special wheel construction of many CBN and diamond wheels and the addition of grinding aids to some wheels, is still generally correct. In this chapter the components of grinding wheels and general guidelines for grinding wheel selection will be discussed.

Types of Abrasives

Abrasive Characteristics

Abrasives are the cutting tools or chip producers in the grinding wheel (see Fig. 3.1). The characteristics of abrasives that determine their efficiency in removing material are crystal hardness, crystal structure, grain shape, the friability or durability of the grain, the chemistry of the abrasive, and whether the grain has been treated or coated.

The hardness of an abrasive grain relative to the hardness of the material to be ground is 1 factor in the ability of a grinding wheel to remove material. In Fig. 3.2 the Knoop hardnesses of various abrasives are shown. Diamond is still the hardest abrasive available, and its hardness makes it extremely efficient in some grinding operations. Other characteristics of diamond, such as its chemistry, make it ineffective in other

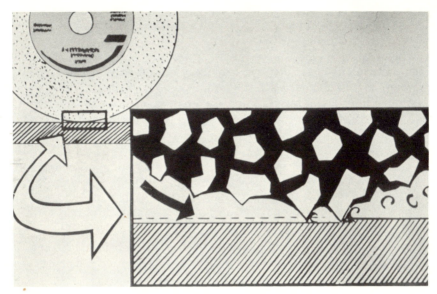

Fig. 3.1   The grinding process

operations. Obviously, however, an abrasive that is significantly harder than the material to be ground will tend to remove more material before it wears out.

The crystal structure of an abrasive grain affects how the grain wears. Abrasive grain that is monocrystalline tends to wear more consistently than an abrasive particle that may have several crystals fused together. Inclusions in abrasive crystals will affect the integrity of the grain. Abrasive grain varies enormously in microstructure.

The shape of an abrasive grain affects both the sharpness of the grain and its wear characteristics. A blocky or spherically shaped grain, when compared with a spindly or jagged shaped grain of otherwise similar

HARDNESS VALUES

MATERIAL	TYPICAL KNOOP HARDNESS NUMBER
DIAMOND	7000
CUBIC BORON NITRIDE	4700
BORON CARBIDE	2800
SILICON CARBIDE	2500
ALUMINUM OXIDE	2000
CEMENTED CARBIDES	1400 - 1800
QUARTZ	800
HARDENED STEEL (ROCKWELL C60)	740
GLASS	300 - 500

Fig. 3.2   Knoop hardness values

properties, will initially need more force to penetrate a material and will fracture or wear more slowly. ✓

The abrasive hardness, crystal structure, and grain shape all contribute to the relative friability or durability of the abrasive. A friable grain is defined as a grain that fractures and exposes new sharp points within the same grain. A durable grain can withstand high forces in the grinding operation without catastropic wear. Those 2 terms are not exact opposites, as some durable grains microfracture to expose new sharp cutting points.

The chemistry of the abrasive affects the ability of the grain to cut in a number of ways. In some wheels (especially vitrified), a chemical bonding occurs between the abrasive and the bond. The chemistry of the abrasive may also affect its ability to grind materials with which it may interact in the locally high-temperature, high-pressure grinding interface. The tendency for diamond to oxidize in high temperatures, for example, as well as its tendency to have low impact resistance, precludes its use for high-temperature, high-force operations.

Covering abrasive grain with chemical, metal, or ceramic coatings can affect the ability of the grain to bond with a particular bond type, can be used to carry heat away from the grinding surface, or can make the grain grind more durably.

Abrasive Types

Abrasives used in the grinding or sanding process can be subdivided into 3 groups: natural abrasives, conventional abrasives, and superabrasives.

Natural abrasives, such as emery, garnet, flint, and iron oxide, are not normally used in grinding wheels. They lack the durability to withstand grinding pressures. Those abrasives are still used in some coated abrasive applications where very light pressures and sharp inexpensive abrasives have utility.

Conventional abrasives are primarily furnaced, fused abrasives, although some nonfused abrasives are available. The 3 primary abrasives are aluminum oxide, silicon carbide, and zirconia alumina.

Aluminum oxide is used in grinding wheels to grind ferrous materials. It is the softest of the conventional abrasives but is relatively impact resistant. There are a number of aluminum oxide grain types the variations of which are chemical purity, grain shape, fused and nonfused, crystal structure, and coatings. The availability of aluminum oxide abrasives varies from a very pure, sharp, friable abrasive to a less pure, blocky shaped, durable abrasive.

Silicon carbide is commonly used for grinding nonferrous materials. It is the hardest of the conventional abrasives but has less impact resistance than aluminum oxide does.

(There are 2 basic types of silicon carbide: black and green.) Black silicon carbide is less pure, slightly more durable, and generally less expensive than green. Green silicon carbide is high in purity, the sharpest conventional abrasive, and a relatively expensive conventional abrasive. Silicon carbide is not normally used in grinding steels, as it is not so efficient (i.e., it does not remove as much material before wearing) as aluminum oxide. Theoretically, that is due to the chemistry of the abrasive and the carbon content of the steels and/or the need for impact resistance when steels are ground.

(Zirconia alumina is used for rough grinding metals, particularly ferrous metals. It has the highest impact resistance of the conventional abrasives. There are 2 types of zirconia alumina: a 15% zirconia alloy and a higher percent zirconia alloy (trade name Norzon). The former is more durable and has high impact resistance. The latter is more friable, with the ability to microfracture and generate new sharp surfaces, and yet is significantly more durable than the aluminum oxides. Zirconia aluminas are not usually used in finishing operations, because they are inefficient under low forces.

The superabrasives are diamond and CBN (Cubic Boron Nitride).

Diamond is available both as a natural, mined abrasive and as a man-made abrasive. The high cost and limited availability of natural diamond has spurred a conversion to man-made or synthetic diamond. Diamond is used to grind carbides, ceramics, glass, and other refractory materials. There are many shapes and purities of diamonds available.

CBN is a man-made abrasive. It is used to grind ferrous materials, particularly hardened steels and alloys. Compared with diamond, CBN is impact resistant, heat-resistant, and chemically less active. As a result, CBN can grind materials that generate a longer chip. However, since diamond is appreciably harder and generally inherently sharper than CBN, diamond is significantly more efficient for grinding materials such as ceramics, which "microchip" when ground.

Nature of Grit Size

The abrasive grain size or grit size is defined in most wheels by a number that correlates to the number of holes per linear inch in a screen that would have holes similar to the average particle size of the abrasive. In other words, a 30-grit particle would pass through a screen having 27 holes per linear inch (See Fig. 3.3). The grit size number is an inverse to the size of the particle (i.e., the higher the number, the smaller the particle).[2]

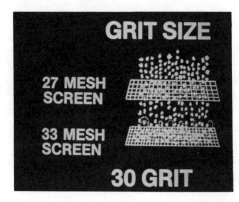

Fig. 3.3   Grit size screening

Coarse grits (having low grit-size numbers) are normally used to rough grind, i.e., remove stock with little regard to finish. Finer grit sizes (having higher grit-size numbers) are normally used where tolerances and finishes are important.

Types of Bonds

Nature of Bonds

Bond is the tool holder of the grinding wheel system. Since most grinding wheels must have the ability to wear to expose new grains to maintain the grinding surface of the wheel, the physical requirements of bonds vary under differing grinding conditions. Bonds vary as to their elasticity or brittleness, durability, ability to withstand heat, and the nature of the bond-abrasive interface. Various bonds also have grinding aids, such as lubricants, added to the bonds to improve grinding performance.

Bond Types

There are 3 major bond families: vitrified, organic, and metal.

Vitrified bonds are essentially glass bonds. They are relatively brittle bonds that rigidly hold the abrasive particle and fracture or wear because of application of force. Vitrified bonds, with comparatively high melting temperatures, are relatively impervious to heat.[3] They usually combine chemically with the abrasive.

Organic bonds include resin, rubber, and shellac. All of those bonds physically surround the abrasive to hold it in the wheel and wear because of heat. All of the organic bonds have some degree of elasticity.

Resin bonds have been developed to do a wide range of jobs. Durable,

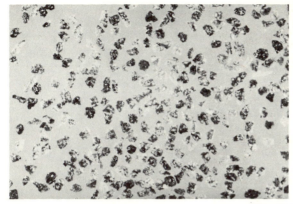

Fig. 3.4   Aluminum oxide grain

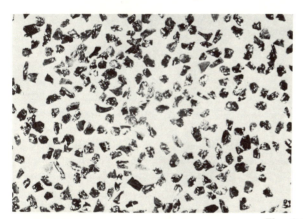

Fig. 3.5   Silicon carbide grain

Fig. 3.6   Norzon grain

impact-resistant, heat-resistant bonds have been developed for rough grinding and heavy stock removal operations. Other resin bonds have been developed to wear under lower unit forces for finishing operations. Grinding aids have been added to resin bonds to add lubricity or to absorb heat away from the grinding surface. Two types of resin are prevalent: phenolic or bakelite resins and epoxy resins. The latter are often described as plastic bonds. Phenolic resins can be used in both wet and dry grinding operations and range from relatively brittle to relatively mild. Epoxy resins should be used wet only and tend to be mild-acting bonds.

Rubber bonds are the most resilient of the organic bonds. They tend to be used where conformability to a surface is preferred.

Shellac bonds are the mildest and most heat sensitive of the organic bonds. They are normally used when grinding a burn-free part while generating a superior finish is the first priority.

Metal bonds are used primarily with superabrasives. They are physical bonds that are more heat resistant than organic bonds and more impact resistant than vitrified bonds. Compared with the fracturing of vitrified bonds or the melting of organic bonds, metal bonds tend to wear attritionaly through the application of abrasive media, either as part of the grinding process or as applied by the operator. Metal bonds are often used in wet grinding applications grinding ceramic, glass, and abrasive materials.

Electroplated metal bonds adhere a single layer of superabrasive onto a metal preform. Electroplating generates a relatively sharp wheel surface, which will tend to cut freely but is not impact resistant and tends to be heat sensitive.

Nature of Grade

Through the life of a grinding wheel, the bond is worn away and the grinding surface is regenerated. That process occurs during the grinding process and, in some cases, is artificially induced by truing and/or dressing. The customer describes the rate of wear of the wheel as its "grinding grade." A hard wheel wears slowly; a soft wheel, more quickly.

A grinding wheel manufacturer defines the grade of the wheel by the amount of bond relative to the amount of abrasive in the wheel. A wheel with more bond per abrasive particle will normally have a stronger or harder bond.

There is frequently a discrepancy between grade as defined by the wheel manufacturer and "grinding grade" as defined by the user. Under the same grinding conditions, a dull, hard wheel may generate heat and vibration, which cause the wheel to break down more quickly than a softer wheel, which stays sharp as it wears. A user may also cause a dull wheel to wear more quickly by removing the excess bond and dull abrasive to resharpen the wheel (see Chapter 4: "Truing and Dressing of Grinding Wheels"). The "grinding grade" or wear rate of the hard wheels therefore becomes softer than that of softer-grade wheels.

There is no industry standard to describe the grade of a grinding wheel. Although most companies designate grade as a letter from A to Z, A being soft and Z being hard, a grade letter does not indicate grade equivalence between 2 different bonds, even from a single manufacturer. Grade letter can be taken as a measure of relative bond strength within a bond system but may not be a good indicator otherwise.

Other Wheel Components

Wheel Structure

Thus far we have defined a grinding wheel as having abrasive particles of a given grit size held by a bond of given hardness. The distance between the abrasive grains is further defined as the wheel structure or porosity. A high-structure number (the number is usually an integer between zero and 25) means more distance between abrasive grains. The optimal porosity for a given operation is defined as a balance between having a high enough structure to allow chip clearance and coolant introduction into the wheel and a low enough structure to ensure enough abrasive particles to do the work required.

As with grade, there may be no relationship between the porosities of 2 bonds having similar structure numbers, as there is no industry standard.

Wheel Treatments

After manufacture, wheels may be treated to aid the grinding process. For example, vitrified wheels may be treated with sulfur for adding lubricity to wet grinding operations or with wax for retarding the loading (impregnation with metal chips) of the wheel when[1] dry grinding aluminum. Treatments are often messy to work with and are used only for special applications.

Grinding Wheel Characteristics

A number of terms are used to define the grinding action of a wheel in relative, and somewhat subjective, terms. They include sharpness, durability, and versatility.

A wheel is described as sharp if it removes a chip quickly at a low enough force level that the part is not excessively thermally damaged. Too sharp a wheel may generate too rough a surface finish on the part. A dull wheel will tend to remove a chip more slowly and/or draw more power and generate a smoother finish (assuming that force levels are low enough to avoid vibration and chatter in the system).

The surface of the wheel is described as open if the wheel is sharp and cutting freely. The surface is conversely described as closed if the wheel is smooth and dull. Opening the wheel face is normally done to promote stock removal; closing the wheel face will usually produce a better finish on the ground part.

A wheel is described as durable if it can withstand high forces over time without catastrophic loss of grinding ability. Durable wheels are important in high-force, high-stock-removal-rate grinding. While durable wheels are hard-grade wheels, they must also have abrasive and bond with high impact resistance.

A versatile wheel is one that grinds a number of parts, materials, and jobs without problems. In job shop situations, such as toolroom grinding, versatility of a wheel can be very important, since wheel changes, excess wheel stocks, and downtime between jobs are expensive.

Specifications of Grinding Wheels

Figs. 3.7 and 3.8 illustrate the common marking system for conventional abrasive grinding wheels and superabrasive grinding wheels. As shown, the pieces of the marking system are generally organized so that the abrasive information precedes the bond and structure information.

The parts of the conventional wheel marking include: abrasive modifi-

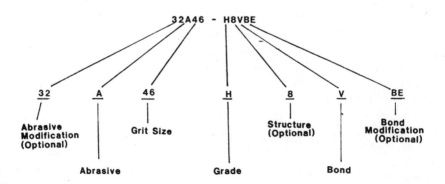

Fig. 3.7   Conventional wheel markings

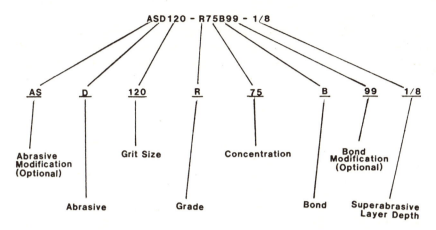

Fig. 3.8   Superabrasive wheel markings

cation, abrasive type, grit size, grit size combination (not shown in Fig. 3.7), grade, structure, bond type, and bond modification. A simple marking needs only to have abrasive type, grit size, grade, and bond type; all other parts of the marking are optional.

The abrasive modification specifies which type of the abrasive is to be used. For example, the abrasive modification would specify for a silicon carbide wheel whether black, green, or a mixture of black and green silicon carbide is to be used. The abrasive modification may be any alphanumeric string, and there are no industry standards.

The abrasive type is generally limited to those previously discussed: aluminum oxide (A), silicon carbide (C), and zirconia alumina (Z)—in superabrasive wheels CBN (B) and diamond (D). The abrasive type may also designate a blend of abrasive types; e.g., AC as a mixture of aluminum oxide and silicon carbide.

Grit size is described, as was previously discussed, by the abrasive screen mesh size. Those sizes differ only slightly and can be considered industry standards.[4] Grit size numbers have been standardized so that 46 is a standard grit size while other numbers between 40 and 50 are not.

A grit size combination number (if shown) designates whether or not a blend of grit sizes has been used. Normally, a 1 affixed to grit size (e.g., 46 becoming 461) designates that no blending has been done. Any other number may designate a blend of grit sizes, and there is no industry standard for grit size combination numbers.

If no grit size combination is shown, the manufacturer supplies a standard combination that may or may not be blended and may depend on product type.

The grade of the wheel, as previously discussed, is normally described as a single alpha character from A to Z, A being soft and Z being hard.

The structure of the wheel (the distance between the abrasive particles) is usually a number between zero and 25. If structure is omitted, the particular bond type has dictated a particular structure, and since the structure is fixed, it is not shown. Not all bonds are available in all structures; in fact, bonds are usually very limited in the available structure range.

The bond type of a wheel is generally designated as a single alpha character: V for vitrified, B for resin, R for rubber, E for shellac, and M for metal.[5] (Superabrasives are not available in rubber or shellac; conventional abrasives are not normally used in metal bonds.)

The bond modification further defines the bond type being used. Similar to the abrasive modification, bond modification is an alphanumeric string, which has no industry standards.

For superabrasive wheels, the markings are essentially the same except for the replacement of structure by the abrasive concentration and the addition of the superabrasive layer depth to the end of the marking.

The concentration of a superabrasive wheel is the amount of abrasive in the abrasive layer. Because of the high cost of superabrasives, a thin layer of grinding wheel may be attached to a core material so that only that outer layer is usable. (Note the abrasive rim construction in superabrasive wheels shown in Fig. 3.9.) The concentration of superabrasive in the layer is designated as a number, usually an integer up to 200. There are no standards for concentration; i.e., a 100 concentration product will vary in abrasive content between suppliers. The concentration of a superabrasive wheel can be considered as linked to the wheel hardness, since the amount of superabrasive in the wheel does affect the wearing characteristics of the wheel.[6] A higher-concentration wheel normally requires higher forces and wears more slowly.

The superabrasive layer depth describes the usable depth of grinding section available. That may be designated as a fraction, a decimal, or, in the case where there is no core, "SOLID."

Principles of Grinding Wheel Selection

Factors to be Considered in the Operation

Seven major factors affect the use of grinding wheels for various operations: the material to be ground and its hardness, the stock to be removed and the finish required, the nature of the coolant (if any), wheelspeed, the area of grinding contact and equivalent diameter, the

Fig. 3.9   Superabrasive wheels

severity of the operation, and the horsepower available to the grind. Each of those factors can affect the selection of the optimal grinding wheel.

Material to be Ground.   There has never been a thoroughly satisfactory method of looking at the properties of a material and using the hardness, tensile strength, or some other parameter to thoroughly describe how that material can best be ground. Knowledge of the hardness, the chemistry, and the nature of chip formation of a material can aid in the selection of a grinding wheel, however.

With few exceptions, superabrasives are used only on hardened materials (in excess of Rc55). When softer materials are ground, the generation of longer chips tends to prematurely degrade the bond or, if the bond is strong enough to resist that wear, the chips may become imbedded in the wheel and actually weld onto the cutting edges. In summary, superabrasives tend to work best on hard and/or brittle materials.

The chemistry of a material affects grinding wheel selection in several ways. Some alloys, such as aluminum, copper, and chromium, contain materials that tend to generate long chips. Other materials, such as titanium, oxidize exothermically. In each case wheel sharpness is crucial, as cooling the grinding operation is essential. Silicon carbide is often used on the above materials because of its inherent sharpness. Depending on the ability to cool the system, soft grades may have to be used to ensure wheel sharpness.

In order to continue to remove stock on a hard material, a grinding wheel must remain sharp. Since the abrasive particle in conventional wheels dulls more quickly when grinding hard materials than when grinding softer materials, soft-grade wheels must be used on hard materials so that new sharp grains are exposed. Harder-grade wheels on hard materials will tend to retain dull grains generating vibration and heat and burn.

Stock Removal/Finish Required.   A simple method of stating this factor is the difference between the part before and after grinding. That includes stock to be removed, dimensions to be held, finish required, and necessary surface integrity

The relationships between those 4 parameters usually need to balance out. In a roughing operation, stock removal rate may be the only parameter of importance, but in finish grinding the other 3 parameters come into play. In order to remove stock without burn, burr, or chatter, a free-cutting, sharp wheel is preferable. However, in order to hold tight dimensions and generate fine finishes, a continuously sharp wheel will tend to lose form and generate poorer finishes than will a duller, harder-acting wheel. Striking the balance between being sharp enough and dull enough is often difficult.

Assuming you have sufficient force to allow a coarse grit to penetrate into a part, coarse soft wheels with friable abrasives tend to cut quickly without surface injury. Fine, hard wheels tend to generate finer finishes and closer tolerances.

Coolants.   Coolants are important regarding their ability to lubricate, conduct heat, and carry swarf from the grinding surface. With refrigerated oil coolants, wheels several grades harder can be used with no metallurgical damage, as compared with doing the same operation dry or with water.

In order to get the most out of coolants, large quantities of clean cool coolant have to be directed at the grinding area. If coolant is directed elsewhere, insufficient, and/or improperly maintained, it may be worse than useless.

High-lubricity coolants also enhance the ability of CBN to cut metal. Straight oil in copious amounts can improve the life of a CBN wheel by a factor of 10 over 10% water soluble. On the other hand, low-concentration coolants tend to work better when one is grinding ceramics or glass. In the first case generating a chip as effortlessly as possible is more important; in the second case cooling the part is the prime consideration.

Wheelspeed.   Before the discussion of wheel speed goes any further, the first rule of grinding wheel safety is *never* to exceed the surface speed caused by rotating the full-sized wheel at the maximum operating speed marked on the wheel. All of the discussion of wheelspeed here is regarding wheelspeeds less than or equal to the maximum operating speed listed on a grinding wheel.

A wheel rotating at high speed is required to take a shallower chip per revolution than is a wheel at slow speed, moving at the same infeed rates. As a result of each abrasive particle being asked to do less work, the abrasive tends to dull more slowly and the wheel tends to wear more slowly. A wheel running at higher speed is said to be harder acting therefore, since it lasts longer. If the wheel is optimized at slow speed, speeding it up may make it too hard, causing burn and chatter.

CBN wheels tend to work better at higher speeds, since they work better when generating many tiny chips than fewer larger ones. Grinding titanium with conventional abrasive wheels may be done at slower speeds so that the chips are less likely to oxidize.

The Area of Grinding Contact.   At a given force level, the area of grinding contact (defined by the equivalent diameter) determines the unit force available to generate a chip. The larger the contact area, the lower the unit force that is available to generate a chip. For large contact areas then, sharp (and relatively soft) wheels are used. For small contact areas, durable abrasives and relatively hard wheels are used to stand up to the higher-unit forces.

Severity of the Operation.   In some precision grinding operations, the grinding cycle is interrupted (e.g., horizontal spindle reciprocating surface grinding when the wheel moves off and then back on the part) in such a way that an additional impact is added to what the grinding wheel must withstand. If that impact is severe, a more durable abrasive and a harder grade is necessary to forestall premature wear. The reverse is true in such operations as creep-feed grinding, where soft, sharp wheels are optimal.

Horsepower.   All other variables being equal and a rigid system being assumed, adding horsepower to a system will tend to cause more heat and force to be applied to the wheel, causing more wear. Removing horsepower from a system may cause the grinding wheel to bounce on the work, since the force available is insufficient to generate a chip, causing friction, chatter, and burn.

It takes more force to generate a thick chip with a coarse grit abrasive than otherwise. As a result, in a low-horsepower system grinding a hard material, fine grits must be used to penetrate the material. For example, for cylindrical grinding Rc60 steels, usually 60-grit and finer wheels are recommemded.

It should be pointed out that the above discussion concerned usable horsepower. Often the available horsepower is not used either because of inefficient systems or finishing operations where more power is deleterious to the grinding process.

Grinding Wheel Safety

It is recommended that all users and potential users obtain and read the ANSI code B7.1, "Safety Requirements for the Use, Care, and Protection of Abrasive Wheels." That document reviews the different shapes of grinding wheels (some of which are shown in Fig. 3.10) and their proper storage, mounting, and use.

Fig. 3.10   Types of bonded abrasive products

Conclusions

The options regarding which grinding wheel to use on a given operation are many: sharp, open, friable, soft, coarse versus dull, closed, durable, hard, fine. They are further clouded by the fact that in most cases a compromise between the extremes is necessary to ensure an acccceptable finished part. System variables can be additive or can cancel each other regarding their effects on the grinding wheel.

In the next chapter, truing and dressing will be discussed. The extent to and manner in which a wheel is trued and dressed can sharpen or dull it.

The final compromise in specifying a grinding wheel is between wheel life and material removal rate. An analysis of the whole grinding system has to be made as to whether the wheel should be designed to last to cut abrasive costs or be designed to remove material quickly to cut labor costs and increase production.

Notes

1. Norzon is a registered tradename of the Norton Company.
2. A notable exception is extremely fine grit product, which may be characterized as a micron size or micron size range.
3. Vitrified wheels, however, cannot be exposed to large temperature variations during grinding, as the bond is not elastic enough to withstand thermal expansion differences within the wheel.
4. There are some small differences between U.S. (ANSI) and European (FEPA) grading. Japanese sizing is dramatically different in fine-grit (numbers larger than 200) product.
5. In superabrasives wheels C may designate electroplated.
6. Some superabrasive products will omit the concentration, in which case the grade letter is used as a measure of both bond and concentration.

CHAPTER 4

Truing and Dressing of Grinding Wheels

William Ault

Introduction

In precision grinding operations, exacting tolerances and finishes often require careful control of the geometry and surface roughness of the wheel. In order to regenerate the wheel face, the wheel is trued and dressed.

Truing is defined as the regeneration of wheel geometry. Dressing is defined as the regeneration of the desired surface condition of the grinding wheel. In some cases truing and dressing is a single operation; in others a wheel must be dressed after truing.

As a grinding wheel is used, it goes through a series of cycles in which the wheel dulls, wears, and new sharp grains are exposed, followed by the next cycle. A wheel may wear unevenly because of unequal forces, unequal stock removal requirements, or to a smaller extent, the random orientation of the abrasive in the bond. The wheel will need to be returned to the desired geometry once the wear pattern becomes sufficiently nonuniform to stretch part tolerances toward unacceptable limits. That is the purpose of truing.

Wheels are dressed for several reasons. They include opening up the wheel face, closing up the wheel face, and removing grinding swarf.

Fig. 4.1 shows a diagram of the face of a superabrasive wheel that has just been trued. Note that the surface of the wheel is essentially smooth without exposed cutting edges or places for chip clearance. Without opening up the wheel face by dressing (shown in Fig. 4.2), the wheel would tend not to penetrate the material being ground. The normal

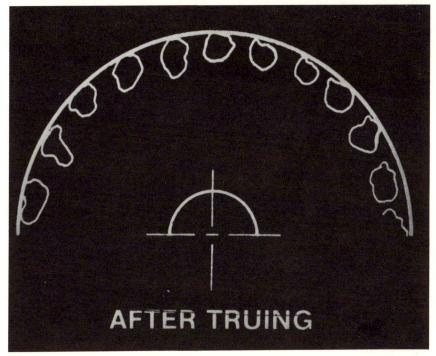

AFTER TRUING

Fig. 4.1   Wheel face profile, after truing

force would tend to vibrate the machine spring, causing bouncing of the wheel on the work. Also the rubbing of the wheel on the part would tend to create excessive frictional heat, which could result in damage to the ground surface. The dressed wheel has exposed cutting edges, and the cleared-away bond allows chip clearance.

In many precision operaions a roughing pass for rapid stock removal is followed by truing and dressing the wheel to ensure part geometry and the required surface finish. In those cases the truing and dressing of the wheel is a single operation with the wheel left in a relatively closed and dull condition to provide a fine finish on the part. Often the wheel is not redressed, but the roughing forces on the next part are used to re-sharpen the wheel for stock removal. The cycle is then repeated for fin-ishing the second part and roughing the third.

Particularly when one is grinding softer materials, which generate longer chips, the grinding face of wheels may become impregnated with swarf. It fills the pores of the wheel and, if not removed, will eventually weld onto the abrasive particles and inhibit the cutting action of the wheel. In that case dressing the wheel will remove the swarf from the cutting surface to resharpen the wheel.

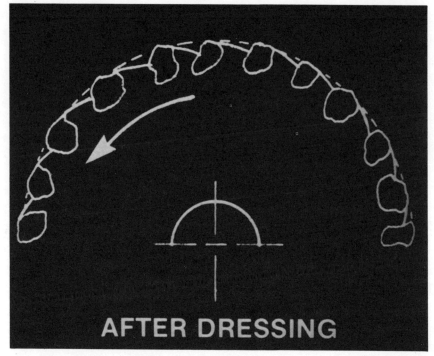

Fig. 4.2   Wheel face profile, after dressing

Types of Truing Tools

Nondiamond Truing Tools

Nondiamond truing tools include abrasive sticks, star cutters, brake-controlled truing devices, soft steel blocks, and crush rolls.

Abrasive sticks are generally conventional abrasive bonded products similar in compositon to grinding wheels.[1] Hard-grade sticks are used to true the wheel to a point such that frictional heat and drag are minimized as the part passes by the wheel during grinding.

Star cutters are used in offhand grinding and some surface grinding operations (such as grinding with cylinder wheels, segments, and disc wheels) as a truing and dressing tool. A star cutter is a free-wheeling set of star-shaped steel blades mounted on a single spindle, which when applied to the wheel crushes the wheel face. It is normally used only with relatively coarse grit conventional abrasive wheels (36 grit and coarser) to create a very rough surface on the trued wheel.

Brake-controlled truing devices are used for truing superabrasive wheels (see Fig. 4.3). The brake-controlled truing device is essentially

Fig. 4.3   Brake controlled truing device

a silicon carbide vitrified wheel applied to the grinding wheel with the 2-wheel axes parallel. The grinding wheel is run at normal speeds, and the braking action of the truing device creates a rubbing action between the grinding and truing wheels. The truing device is traversed through a plane perpendicular to the side of the wheel when one is straight-face truing. The brake-controlled truing device is seldom used for form truing.

Another method for truing resin-bonded superabrasive wheels, particularly on surface grinding operations, is grinding a soft steel block. The heat involved in grinding the soft material trends to cause the wheel to wear back to a round condition. That truing method is relatively traumatic to the bond on the wheel face, and while it is effective, it is not recommended.

A nondiamond method for truing intricate forms into wheels is crushing truing. A powered crush roll with the desired form is applied to the free-wheeling grinding wheel, and the wheel is crushed to shape as it rotates. The crush roll may be cast iron, steel, or carbide depending on the intricacy of the form, the surface to be generated on the wheel, and the type of wheel being crushed. That truing method is particularly effective on vitrified bonded wheels and generates a relatively open wheel face.

Diamond Truing Tools

There are basically 3 types of diamond truing tools: shank tools, diamond blocks, and rotary diamond cutters.

Diamond Shank Tools.   Diamond shank tools are divided into 3 types: single-point, cluster, and multi-point nibs. Shank tools are made up of 3 parts: diamond, bonding material, and the shank. The diamond in each type of tool can vary depending on the use of the tool. The bonding material is usually a metal bond matrix which, like metal-bond grinding wheels, creates a physical bond to hold the diamonds in place. The shank is dependent on the ability of the machine to present the tool to the wheel and varies in diameter (or rectangular dimensions in some cases), length, and the angle at which the diamond-bonded section is held to address the grinding wheel.

Single-point and cluster tools use primarily natural diamond crystals.[2] Those diamonds tend to vary widely in shape, quality, and size depending on the type of tool, the wheel to be trued, the manufacturer of the tool, and the ability of the customer to fully utilize a tool. (Some of those tools are shown in Fig. 4.4.)

Most single-point tools can be purchased with resettable diamond points. Depending on the size, shape, and quality of the diamond, 1 to several new sharp points (previously held in the bond matrix) can be exposed when the tool is returned to the supplier so that the diamond can be reset in the matrix. Should it be difficult to ensure removal of the diamond tool before either fracturing of the diamond or loss of the diamond from the bond, nonresettable tools (throwaways) are also available.

Fig. 4.4   Single-point and cluster diamond tools

There are no industry standards for describing the quality of diamonds in a tool. While diamonds are judged on size, shape, and clarity, the quality of diamonds used in truing tools varies dramatically and is directly related to diamond cost. A high-quality single-point diamond tool with a guarantee of a certain number of resets may be a factor of 10 (or more) more expensive than a throwaway tool. A quality diamond will be economically superior in the long run if the truing environment is correctly maintained and the tools rotated and replaced as necessary.

When one is truing wheels with single-point tools, the diamond tends to wear flats on the crystal. As those flats are generated, the frictional drag on the diamond is increased, which causes more heat and subsequently accelerates the wear on the diamond. It is important, therefore, that the diamond be rotated periodically so that a new sharp edge is turned to the wheel.[3] Obviously, that works only if the diamond addresses the wheel so that the major axis of the diamond is not perpendicular to the wheel face. Usually single-point diamonds address the wheel from 15 to 30 degrees off the perpendicular with the rotation of the wheel away from the shank to reduce impact on the diamond.

The size of the diamond in a single-point tool is dependent on the radius to be dressed in the wheel and the size of the wheel.

Single-point tools are not recommended for straight-face truing of wheels, as other tools are usually economically superior. For truing tight radii, smaller diamonds are often used for sharpness and the ability to hold the diamond in the small amount of bond available on small tools. For large-diameter wheels, large diamonds are recommended, because they can be more firmly held to resist impact and usually last longer on the larger volumes of grinding wheel that are usually required to be trued off. With the geometric rise in cost of diamond versus diamond carat weight, it has become difficult to economically justify diamonds larger than $1/2$ carat.

Cluster tools are essentially multiple single-point tools. The diamonds are carefully hand set into the bond matrix so that a properly used tool will have several diamonds contacting the wheel similarly during truing. They are used in straight-faced truing and some very limited form truing operations where additional diamonds can accelerate the truing of a wheel. The diamonds used in those tools tend to be smaller than those used in single-point tools and are normally not resettable. These tools may not be rotated during their life cycle. They may have multiple layers of diamonds (not more than 2 or 3), which are exposed as the tool wears.

Multi-point diamond nibs have diamond chips relatively randomly spaced in the bond matrix. Much as a grinding wheel, those tools were, constantly exposing new sharp diamonds while shedding dull, worn crystals. The tools are used for straight-faced dressing. They do not have

Fig. 4.5   Typical multi-point diamond nib

to be rotated and usually have superior life to single-point and cluster tools. (A typical multi-point nib is shown in Fig. 4.5.)

Diamond Blocks.   Diamond blocks are a specialty diamond tool used primarily for formed reciprocal table surface grinding applications. The block is similar in design to either a cluster or a multi-point nib.

Rotary Diamond Cutters.   Rotary diamond cutters are mounted on driven spindles and advanced toward the grinding wheel so that the negative of the cutter shape is imparted to the grinding wheel. The axis of the cutter spindle is either in the same plane as the axis of the grinding wheel, or in the case of cup-shaped cutters, the cutter spindle axis is perpendicular to the wheel axis. Rotary cutters can be used for truing either straight or formed wheel faces.

There are 3 types of rotary diamond cutters: hand set, random set, and reverse plated. In hand-set cutters, a single layer of diamonds is painstakingly set by hand so that the pattern of diamonds is very uniform and the sharpest points are exposed. The diamonds are then set in a metal bonded matrix. Random set cutters are similar to metal bond diamond grinding wheels except that normally a single layer is metal bonded onto the steel preform. Normally that method is not used for intricate forms. Reverse plated cutters are usually manufactured by hand setting the diamonds on a form that is the negative of the finished cutter and then plating the diamonds onto the preform, using the negative as a type of mold. The diamond surface is made up of the mounted flats so that the reverse plated cutter can be very precisely manufactured without grinding the cutter after molding and the cutter is durable and uniform. In summary, random set diamonds are used primarily for straight-face truing, hand set diamonds for simple forms, and reverse plated diamond cutters for intricate forms to tight tolerances.

Truing with Diamond Truing Tools.   As previously discussed, the nature of diamond is that it is susceptible to heat and has low impact resistance. As a result, truing with diamond tools should be done, if possible, with low-impact forces and in flood coolant. (That last recommendation runs contrary to popular practice, which is to use spray mist or no coolant. Historically that has been done to help the operator hear the operation or to create a rougher dressing of the wheel with the truing tool. Both the wheels and the diamond tools will work more efficiently if flood coolant is used.)

Fig. 4.6 shows a chisel-type single point, which is being used to form true a creepfeed wheel. (The wheel guard has been opened to show the wheel and tool and should be closed before operating the machine.) Note the additional coolant nozzle immediately above the diamond tool to ensure coolant flow at the point of contact.

Types of Dressing

Nondiamond Dressing Tools

Nondiamond dressing tools include abrasive sticks, star cutters, and grinding aids. The abrasive sticks used as dressing tools tend to be softer

Fig. 4.6   Form truing operation using single-point diamond tool

and finer than the wheel being dressed. Star cutters (again used only for roughing and rough surfacing operations) are as much a dressing tool as a truing tool. Grinding aids, such as wax or rosin, are used, often in stick form, to retard loading when grinding soft stringy materials.

Diamond Dressing Tools

Diamond truing tools are often used simultaneously as dressing tools.

Shank tools, which are traversed past the grinding wheel, can dress a wheel closed or open depending on dress depth and dress lead. The dress depth, the infeed of the tool per pass, can vary from .00005 inches for extremely fine finishing operations to .002 inches on roughing operations (heavier dress depths are not recommended). The dress lead (the traverse distance per wheel revolution) defines the "threading" pattern on the face of the wheel. Those 2 factors can combine to generate an open face (deeper dress, faster traverse) or a closed face (shallower dress, slower traverse).

Fig. 4.7   Typical shank tools for O.D. grinding

The rotational speed of a rotary diamond cutter relative to the direction and speed of the grinding wheel will also affect the dressing action. If the grinding wheel is run in the opposite direction to the cutter at the point of contact, the shearing action involved creates a very dull closed wheel. If the grinding wheel and truing cutter are run in the same direction at the same speed, a type of crushing occurs that opens the wheel face. If the truing cutter is then slowed to roughly 80% of the grinding wheelspeed, the resultant drag tends to create a sharp, open wheel face without the loose abrasive being crushed into the wheel.

Conclusions

As will be discussed in a later chapter, the truing and dressing of super-abrasive wheels is a critical problem in maintaining a constant cut rate and finish and in justifying the expense of the wheel and maintaining the grinding system. The truing and dressing of conventional abrasive wheels mechanically corrects parameter inconsistencies, regenerates tolerances and finishes, and ensures a controlled system. The importance of picking the correct tool, applying it in the correct manner, and maintaining or replacing the tool is crucial to most precision grinding operations.

Notes

1.   One notable exception to that rule is a centered boron carbide (Norton trademark Norbide) which is used for offhand truing of tool-grinding wheels.

2.   Man-made single diamond crystals have been limited in size, although that may change. Some tools are also available with a polycrystalline manufactured diamond on a tungsten carbide substrate. Those tools tend to wear evenly and have high impact resistance but may not be so sharp as natural diamond.

3.   Some special tools, such as chisel points, may not be rotatable and should be replaced when flats occur.

CHAPTER 5

Grinding with Superabrasives

R. L. Mahar
Norton Company

Introduction

The term "superabrasives" is used to differentiate diamond and cubic boron nitride (CBN hereafter) abrasives from the "conventional" precision grinding abrasives silicon carbide (SiC) and aluminum oxide (Al_2O_3). The key property of an abrasive distinguishing the "superabrasive" from the conventional abrasive is hardness, which generally translates to cutting edge durability. Other properties that are important are impact strength, chemical activity, and thermal stability.

Diamond is the hardest material known and is used extensively in grinding tungsten carbide and nonmetallics, such as stone, concrete, ceramics, and glass. Diamond has some utility on ferrous materials, but its chemical activity generally limits its performance. Ferrous materials have an affinity for carbon, and at grinding temperatures that causes a chemical erosion of the diamond abrasive particle, since it is made primarily of carbon. The cutting edges round, causing a dull cutting edge, causes frictional rubbing and inefficient chip formation. That requires higher grinding forces, which risks metallurgical damage to the part being ground. Therefore diamond is not generally the superabrasive of choice on ferrous-based materials.

CBN is made of boron and nitrogen, which avoids the carbon affinity problem in grinding ferrous materials. It is the second-hardest material known, does not occur naturally, and is 2 to 3 times harder than aluminum oxide abrasive. Because of its chemical inertness, it is the superabrasive of choice on ferrous alloys as well as nickel and cobalt-based

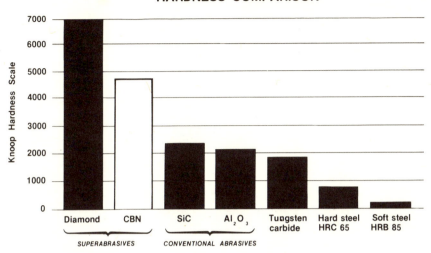

Fig. 5.1   Abrasive hardness comparison bar chart

alloys. CBN is synthesized at high temperature and high pressure in the same process used to make synthetic diamond.

CBN was first synthesized in 1958[2] but not commercialized until 1969. It was readily adopted for high-speed steel-cutting tool resharpening, but penetration into production-grinding applications has taken considerably longer because it is technically more complex to apply than conventional abrasives. However, the use of CBN in production grinding is developing rapidly in Japan, Europe, and the United States. Consumption of CBN grinding wheels and hones used for production grinding already significantly exceeds that used for tool maintenance (resharpening).

CBN is frequently referred to as the "abrasive of the future"; however, that should not be interpreted to mean it will obsolete conventional abrasives. CBN is believed to be economically viable on an estimated 25% of the precision grinding market, but its actual usage is far below that and growing rapidly. As CBN abrasive costs come down and part materials are upgraded (i.e., become more difficult to fabricate), the potential market for CBN could increase even more.

Since the focus of this text is precision grinding of ferrous materials (steels, etc.), the remainder of the chapter will concentrate on the benefits, use strategy, conditions for success, and applications of CBN. For more information on the grinding of carbide and nonmetallics, you may refer to The Diamond Wheel Manufacturers Institute series of articles[3] listed in the bibliography.

MISSED OUT

Selection Criteria: CBN vs. Conventional Abrasives

CBN should be considered whenever conventional abrasives have difficulty achieving a satisfactory work result in terms of economics (total grinding cost per part), productivity (efficient use of machinery and labor), part quality (dimensional and geometric), and metallurgy (surface quality) or whenever automated production lines requiring minimal supervision and labor content are contemplated.

CBN wheels are being successfully used in a variety of precision grinding applications: internal, external, cam-contour, centerless, surface, double-disc, creep feed, endmill feature, form, fluting, slot/groove, thread, jig, and honing.

CBN is very nearly an ideal precision abrasive. Its hardness and cutting edge durability provide long-lasting efficient cutting capability. Consequently, when applied properly, CBN will easily grind essentially any ferrous alloy as well as nickel and cobalt-based superalloys. Therefore, technical ability of CBN to grind a material is almost never the issue. Economic criteria govern the decision.

Economics are almost totally dependent on the proper application of CBN through competent system and application engineering. CBN is more technically complex to apply than conventional abrasives because of the "use strategy" that must be adopted (see below).

Conventional abrasive grinding systems are more forgiving and adaptable than CBN, which can be an advantage depending on the needs of the job. Each situation must be analyzed to establish the end result needs and restrictions.

CBN is the abrasive of choice for hard and tough-to-grind materials, such as M and T series high-speed tool steels and high nickel superalloys. Conventional abrasives have difficulty cutting those materials. CBN should be considered for any hardened ferrous material Rockwell C 50 and harder. Softer steels (as low as R_c 22) can be rough ground with CBN at high metal removal rates (up to 4 cubic inches per minute per inch of wheel width) under specific conditions, which may be economically viable.

CBN should also be considered where part geometry and dimensional tolerances are extremely tight, even though the material hardness is not a problem for conventional abrasives. The trend today is to significantly improve the quality of parts to assure reliable and long life in their functional use.

Because a CBN grinding wheel surface wears very slowly compared with conventional abrasive wheels, part quality in terms of roundness, straightness, flatness, and dimensional consistency is often more easily achieved with CBN.

Finally CBN should be considered where surface integrity of the part must be assured, i.e., freedom from metallurgical burn caused by grinding. CBN's ability to grind efficiently (with low grinding forces because the CBN abrasive grain cutting edges are not dulling rapidly) is also more likely to produce beneficial compressive stresses instead of tensional stresses, which can cause premature surface failure.

What is Driving CBN Technology?

The need for:
(1). Improved productivity and lower total costs per part.
(2). Tighter part quality and tolerances
- better part geometry (roundess, flatness, etc.)
- tighter dimensional tolerances
- improved surface integrity
- tighter statistical quality control limits
(3). Upgrading part materials
- take advantage of superior materials, which may have been bypassed because they are difficult to fabricate.
(4). Factory automation
- reduce labor content leading to full automation
- improve asset utilization
- permit automation with confidence

If we are to be competitive in world markets, we must make significant progress in those areas. Manufacturers who ignore those needs and trends will have difficulty competing profitably.

CBN Benefits

The major benefits CBN can deliver are:
(1). **Higher Productivity.** More "good" parts per machine, per labor hour, per capital dollar, etc.
(2). **Lower "Total" Costs per Part.** CBN when properly applied can reduce total grinding costs from 25% to 50% or more. The emphasis here is on "total" cost because the "abrasive wheel cost" per part will most likely increase, but that is easily offset by larger labor and overhead-costs-per-part decreases resulting in major "total" cost savings.

CBN can sometimes be justified at equal or higher total grinding costs per part versus conventional abrasives because of significant savings "down the line," such as reduced inspection costs or simplifying mating part classification and sorting because of tighter size distribution with CBN ground parts.

Specific CBN benefits are:

(1). **Shorter Grinding Cycles.** The goal is to achieve the part geometry (straightness, flatness, roundness) required. That is because CBN cuts any ferrous-(iron-) based or nickel-based material *very efficiently* and *easily*. CBN does not dull quickly like conventional abrasives, so forces stay low and cutting efficiency high—there is no need to spend time continually truing/dressing (as with aluminum oxide) to "keep it sharp" and avoid all the grinding problems a "dull" wheel causes.

(2). **Avoidance of Grinding "Burn" (Metallurgical Damage).** Burn is caused by "dull" abrasives, which cause high frictional heat and high grinding forces requiring high grinding energy. CBN's superior harness (2½ times aluminum oxide) and resultant cutting edge durability (sharpness) avoid those common causes of "grinding burn."

(3). **Consistent Part Quality (Size, Roundess, Straightness, etc.)** CBN's inherent efficient cutting and slow-wearing characteristics make the grind very consistent part after part, which results in: (a) Less operator attention required at the machine for checking parts and making adjustments, (b) fewer rejects or reworks, and (c) less inspection time.

(4). **Difficult Materials Ground Easily.** CBN's superior hardness permits it to *easily* cut the hardest of ferrous-(iron-) and nickel-based materials. But material hardness is not a limiting use for CBN. On materials as low as $R_c 50$, where tighter tolerances are required, a slow-wearing CBN wheel can prove economically viable.

(5). **Tighter Part Tolerances.** CBN's slow-wearing characteristics together with excellent machine accuracy team up to produce very tight part tolerances. Process engineers are increasingly demanding tighter part tolerances, which CBN can deliver. That can avoid costly part inspection and classification operations down the line (often required to assure proper fit and part function in assemblies).

(6). **Fewer Wheel Changes More Machine Uptime.** Especially useful in small-diameter wheel (less than 1-inch diameter—usually internal grinding) situations. In 1 major I.D. operation the machine is down for 5 minutes out of every 25 with conventional wheels (200 parts/wheel). The CBN wheel lasts a week or more (more than 25,000 pieces), giving a wheel change cost/part of $.0003 vs. $.018 for the aluminum oxide wheel.

(7). **Less Labor—One Operator Tends More Machines.** Because CBN wears so slowly and makes consistent parts, the operator does not have to "baby-sit" the machine, constantly checking part quality, making adjustments, and changing wheels. That means 1 operator can tend many more machines, significantly stretching the "labor" dollar spent.

(8). **Machine (Asset/Capital) Utilization.** With CBN more "good" parts can be produced per hour requiring fewer machines to get the "production" out. If new machines are being considered, that means fewer are required. If machines exist, the "excess capacity" created by using CBN can (a) be sold, thus freeing capital dollars and floor space, or (b) "held" until additional capacity is required, with purchase of new machines avoided to accommodate the "growth" needs.

(9). **Lower Diamond Truing Costs.** Frequent truing/dressing with diamond truing tools (single point, multipoint, and rotary cutters) to keep conventional wheels sharp and cutting efficiently is a high cost of grinding. The CBN crystals' inherent and long-lived cutting edge sharpness eliminates that reason for continual truing and the high costs associated with it.

Conditions for Success with CBN

Successful application of CBN is not a mystery. It requires an understanding of the fundamentals for success, a commitment to make CBN work, and the resources for successful implementation. If those conditions are met, the probability of success is very high.

The CBN "use strategy" is 180 degrees opposite that of conventional abrasives (aluminum oxide or silicon carbide). The grinding system when using conventional abrasives is fairly forgiving in that system faults or weaknesses can generally be overcome by "dressing" the grinding wheel with a diamond tool after every part or several times per part. That keeps the grinding wheel geometrically correct and sharp. Conventional grinding wheels, being relatively inexpensive, can be treated in that manner.

CBN wheels, because of their high cost and the need to ensure economic results, must be treated exactly the opposite—that is, the grinding cycle and grinding system must be "balanced" (or stabilized) so that many parts can be produced before ever having to touch the CBN wheel with a truing or dressing tool. Typically 25 to 100 parts must be ground between truings of the CBN wheel. That can be achieved with competent application and system engineering provided that the grinding system components are sound.

CBN is an excellent cutting tool. Failure to attain expected results is rarely a fault of the abrasive, but rather almost always is due to deficiencies in machinery, wheel selection, wheel preparation, grinding parameter selection, commitment, or inconsistent incoming part conditions.

The following "Conditions for Success" will generally assure success with CBN.

(1). **A Well-Defined Problem to Solve.** The problem must be clearly in mind and expressed in terms of low productivity, high part costs, tight tolerances, too high rejections, metallurgical damage, difficult material, machine downtime (frequent wheel changes), long cycle times, etc.—i.e., all the items the "CBN Benefits" address. Lower abrasive cost/ part is not a likely benefit from CBN; it will probably increase but provide large savings in labor and overhead costs, causing total costs to go down significantly.

(2). **Commitment to Work with CBN.** Although the application and system engineering knowledge is building rapidly, most production applications require some changes and system adjustments. The fundamental success elements are known, and the end user must be flexible, adaptable, and willing to provide some of his own resources to make CBN work. Changes may be necessary in coolant type or the coolant delivery system. A spindle may need rebuilding or pulleys changed to increase grinding speed, or special fixturing made to adopt truing-dressing systems. Every situation is different and must be assessed systematically. Most production jobs require some amount of custom engineering.

(3). **Commitment to "Total Cost" Accounting—a Must.** Initial CBN wheel acquisition costs may be 10,100 or even 200 times that of a conventional abrasive wheel. Even though the CBN wheel will grind many more parts per wheel—up to 60,000; when the "abrasive cost"/part is calculated, it will probably average 2 times conventional abrasives— sometimes 3, 5, or 10 times, but occasionally less than conventional abrasives. Obviously, that is not where the CBN benefits come from. The savings, 25% to 50% of total grinding cost (see Fig. 5.2), come from lower labor and overhead grinding costs, fewer wheel changes, faster cycle times, fewer rejects, lower inspection costs, one operator running more machines, higher asset utilization (more up time), etc.

Factory managers and process, manufacturing, and industrial engineers understand "total cost" concepts. Their commitment to making CBN work and adopting "total cost accounting" is an absolute necessity to justify CBN economically and improve their cost position.

(4). **An Adequate "System" to Use CBN Effectively.** "System" = machine/wheel spindle/drive, part holding, fluid delivery, truing, dressing.

The machine does not have to be "new" or "designed for CBN" but must be basically of rigid design, well maintained, with very low vibration characteristics. A machine designed for CBN, such as a Toyoda CNC CBN camshaft contour grinder or a Huffman CNC endmill

"TOTAL" COST ACCOUNTING

$$\begin{array}{c} \text{ABRASIVE} \\ \text{COST/PART} \end{array} + \begin{array}{c} \text{LABOR OVERHEAD} \\ \text{COST/PART} \end{array} = \begin{array}{c} \text{"TOTAL"} \\ \text{COST/PART} \end{array}$$

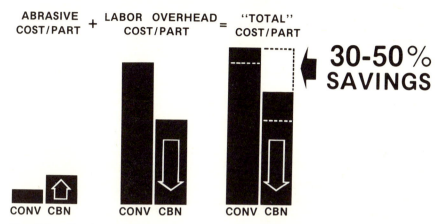

30-50%
SAVINGS

CONV CBN   CONV CBN   CONV CBN

Fig. 5.2   Total cost accounting

grinder, gets maximum utility from CBN. However, an adequate machine can still deliver significant CBN benefits.

The machine *must have* capability of accurate and repeatable infeed of 1/10 of 1/1000 inch (.0001 inch/.0024mm) radially so that truing can be accomplished technically and economically. Infeed capability of 50 millionths inch (.000050 inch/.0021mm) is ideal. Also "skip dressing" capability is a must.

The part holding system must be rigid, positive, well maintained, and free from backlash or vibration.

For fluid type/delivery straight oil is generally the best. The next best choice is a 10% solution of heavy-duty water-soluble oil. On large-area contact operations (double disc, etc) a 2% to 4% concentration of medium-duty water soluble oil has been satisfactory.

Optimum fluid delivery systems have 120 psi (8–9 Bar) and higher with a volume of 30 to 50 gallons/minute (127 to 211 liters/min). They are specifically designed to achieve "optimum" results from CBN. The general rule is to get pressure and volume as high as possible. How much is "enough" is not known with certainty and varies with each application. End users must be willing to make changes if necessary. The *fluid delivery is very important to achieve satisfactory CBN economics*. Proper fluid delivery reduces bond/abrasive wear and reduces frictional heat during grinding.

Truing and dressing are very important. The proper initial preparation of the CBN wheel is an *absolute necessity*, as is the periodic maintenance throughout its life. "Truing" refers to making the CBN wheel round and

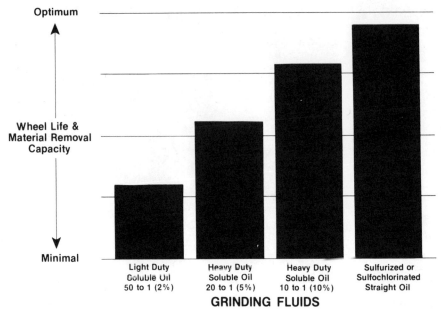

Fig. 5.3   Effect of grinding fluids

running "true" or concentric with its axis of rotation. In addition, it imparts the desired geometry to the CBN wheel face, such as straightness or a form. Initial truing can be minimized when one is mounting a CBN wheel by taking care to get the wheel running as concentrically as possible by using a dial indicator and lightly tapping the wheel into truth with the wheel flanges snug before final tightening.

Since CBN wheels are very expensive compared with their conventional abrasive counterpart, economic viability dictates maximizing the number of parts between retruings. Truing by design always results in a loss of expensive CBN abrasive particles from the wheels' surface. Application engineers always strive to minimize that loss.

The "truer" (rounder) the CBN wheel is, the higher the number of satisfactory parts it will grind before again having to be "retrued" to reestablish wheel geometry (flatness, straightness, roundess, or form) and finish (i.e., cutting down those exposed grains that protrude above the rest). CBN wheel roundness to within .0001 inches (.0025mm) is considered satisfactory; however, roundess to 1 micron (.000040 inches) or less will extend the number of parts between truings. System vibration, whether from poor spindle condition, poor wheel balance, out-of-balance machine components, including motors and drive systems, or external vibration, will all cause premature deterioration of the CBN wheel face, which causes poor part geometry and surface finish, necessitating more frequent retruings, wheel loss, and poor grinding economics.

Truing can be achieved with a 60/80 grit-size multiparticle diamond nib. Some vitrified bonds can be trued with single-point diamonds. However, powered rotary diamond truing tools are recommended and preferred to achieve easy, consistent, and effective truing of CBN wheels. Rotary diamond truers require the lowest normal forces of all truing methods, and low truing forces are necessary for maximum CBN wheel roundess.

The truing tool should be traversed across the CBN wheel at a depth of .0001 inch (.0025mm) or less and at a traverse rate of 20 inches/minute or faster. Larger depths of infeed worsen the economics and increase forces risking springing of the truing tool away from the wheel, causing lobing and poor roundness. System rigidity dictates what level of truing forces are tolerable for one to still get a satisfactorily round wheel.

Conventional abrasive wheels true more easily than CBN wheels, and most truing mechanisms on machines are designed with sufficient rigidity to true conventional wheels. CBN truing forces have been measured, and they approximate truing forces of conventional abrasive wheel truing at truing infeeds 1/10 of those used for conventional wheels.[4]

After a wheel is "trued," it is generally very smooth and dull (especially resin or metal-bonded CBN wheels). The wheel now requires "dressing." CBN particles are in the same surface plane as the bond matrix that holds them. The trued wheel now needs to be "dressed", a procedure that erodes the bond matrix from around the CBN particles,

Fig. 5.4   Truing & dressing

CBN WHEEL AFTER TRUING AND DRESSING

AFTER TRUING   AFTER DRESSING

thus exposing the CBN particle and allowing it to penetrate the workpiece and create a chip.

Dressing also allows space between the CBN particles for grinding swarf (chips) to be carried out of the cutting zone. The bonding matrix should be relieved approximately 25% to 30% below the tip of the CBN particle. That permits strong bond anchoring of the expensive CBN particle. "Overdressing" will cause premature CBN particle loss (called "pullout").

Dressing can be accomplished with a hand-held soft fine vitrified bond (220-H) abrasive stick. However, considerable technique and judgment is required to do it effectively and consistently. That method is not suitable for high production systems. A controllable dressing system, such as an abrasive slurry or an air jet with loose abrasive or a fixed-feed/fixed-force abrasive stick feeding mechanism, is necessary to achieve timely and consistent dressing results. Such dressing units are available commercially.

Machines like the Toyoda CBN CNC Camshaft Contour (Lobe) grinder have been specifically designed for CBN wheels and have very effective automatic truing and dressing systems.

Considerable space has been devoted to truing and dressing because they are *critical* elements in the successful application of CBN wheels.

(5). **Freedom to Change Grinding/System Parameters.** CBN is technically advanced and requires solid application and system engineering usually on site. Generally several elements of the system and the grinding process parameters must be changed. The end user must be willing to make changes as guided by application engineers.

(6). **Part Condition from Previous Operation.** Parts coming to the CBN grinding operation must be consistent (size, stock left for removal, shape, roundess, etc.), which means the previous operation must be in good control. If it is not, it could cause difficulty in getting consistent quality from the CBN operation. It is a case of quality in, better quality out!

Wheel Design

CBN abrasive costs several thousands of dollars per pound compared with conventional abrasives, which rarely cost more than 1 dollar per pound. Consequently CBN wheels are generally designed with a $\frac{1}{16}$ to $\frac{1}{8}$ inch (3mm) radial depth of CBN bearing material. The remainder of the wheel (the "core") is made of inexpensive aluminum, steel, or plastic. CBN depths can be larger than $\frac{1}{8}$ inch (3mm) if necessary. Although

CROSS SECTION OF WHEEL DESIGNS

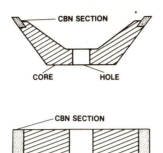

Fig. 5.5   CBN wheel designs

there are many standard shapes of CBN wheels, it is best to consult your grinding wheel supplier to optimize the wheel shape for the specific application.

Wheel Operating Speeds

While Norton CBN wheels will grind very effectively at conventional speeds (in the range of 5,000 to 6,500 feet per minute), their efficiencies may be improved on many materials as speeds are increased. Several machines designed for CBN operate around 9500 FPM (47 meters/second).

At higher speeds wheel life is improved, and it is easier to obtain the desired workpiece finish, provided that system vibration is adequately controlled. Perhaps most important, stock removal capability increases. The only limitations to speed are the wheel type and design, wheel/spindle balance, and the capability of the machine system.

Standard maximum operating speeds mandated by the U.S.A. ANSI Safety Code are 6,500 SFPM for vitrified-bond wheels, 9,500 SFPM for resinoid-bond wheels, 12,000 SFPM for metal-bond wheels, and 16,000 for steel centered cutoff wheels. Speeds higher than those may be approved under conditions specified in Section 8 of ANSI 07.1 Safety Requirements.

"Faster is better" is not an absolute rule, however, because the specific material ground and machine system must be taken into account. Speeds between 5,000 and 7,000 SFPM have been found very satisfactory on the refractory high nickel alloys, such as Inconel. A reminder: *Never run any grinding wheel beyond the manufacturer's limits clearly marked on the wheel.*

Bond Selection Criteria

There are 4 basic choices for the bonding medium that holds the CBN abrasive. They are resin, metal, vitrified, and plated. Each has its advantages and disadvantages, which means the choice of bond must be matched with the application requirements of the job. Grinding wheel supplier application engineers and technical sales personnel are trained to specify the proper bond.

Resin bonds are the most popular because they are the easiest to apply and to true and dress. They impart a reflective finish to the workpiece. There is a broad availability of shapes and sizes. Resin bonds can be used wet or dry. They are generally the first choice for many applications. Their drawbacks are that they are affected by excessive grinding heat (causes bond deterioration and loss of CBN grit retention) and are limited in their mechanical holding strength of the CBN particle, which limits their metal-removal rate capability.

Vitrified bonds are not affected by grinding heat. They are more difficult to apply generally and because they are brittle (a glasslike matrix) and can break if mistreated. There is little "give" to vitrified bonds (unlike resin), which makes them more unforgiving. Vitrified bonds are mechanically strong and durable. They should be used only with a grinding fluid, and they grind at low forces, which makes their cutting efficiency superior to that of resin and metal bonds. Properly selected and applied, vitrified CBN wheels are very productive.

Vitrified bonds have small pores that help get grinding fluid into the cut and chips out. They generally create a matt finish. Truing and dressing generally occur simultaneously, and they can be single-point diamond trued (especially small—less than 2-inch diameter wheels); however, rotary-powered diamond truing is always recommended. Shape and size availability is more restricted than with resin bonds. Vitrified bonds are the bonds of choice for internal, crush-true, and creep-feed grinding applications.

Metal bonds are the strongest and most durable of the bond systems. They generally require higher power because of the metal bond friction on the metal part. Metal bonds take longer to true and dress but hold form well because of their durability.

Rotary diamond truing is generally recommended, although they can be trued with multi-particle stationery diamond truing tools. Metal bonds are first choice in production honing, slotting/grooving applications, and high precision cutoff.

Plated bonds are usually nickel and consist of a single layer of CBN attached by an electroplating process to the surface of a precision-machined steel form. When that layer of CBN is used up, the wheel must be discarded or replated. Truing and dressing are not required. Plated wheels are very aggressive and capable of high metal removal rates at low specific grinding energy. Surface finish is usually quite rough—higher than 32 microinches. Finer finishes require trading off metal removal capability. Shape, form, and size availability are extensive. Plated wheels are used very successfully for grinding forms in high nickel alloys, such as Inconel, etc., and for jig, slot/groove, and general form grinding.

CBN Principles of Grinding

The "Principles of Grinding" listed in Chapter 2 also apply to CBN[5]. CBN wheels behave like all grinding wheels except that the numbers of the work removal parameter, forces, and G-ratios, etc., are different. CBN wheels dull very slowly, thus adding a "time function" to the CBN wheel characterization.

Fig. 5.6 shows the effect of wheel speed on the grinding performance,

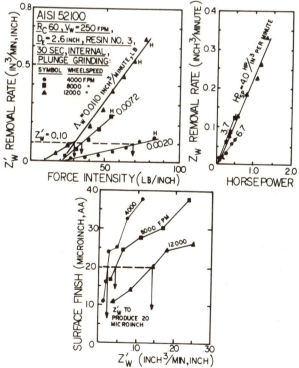

Fig. 5.6   Effect of wheel speed on grinding performance

as measured by WRP, the Work Removal Parameter, and P'_{sp} the specific power. On the left graph, plotting the Z'_n metal removal rate (inch³/minute, inch of width) versus the normal force intensity between the wheel and work (lb/inch of width) gave 3 distinct curves whose slopes have the units "cubic inches being removed per minute per pound of normal force." That is the work removal parameter, WRP. Increasing the wheel speed from 4,000, 8,000 to 12,000 f.p.m. (feet per minute) improved WRP from 0.0020, 0.0072, to 0.0110 inch3/minute, lb. The significance of that was that to grind at some metal removal rates, Z'_w, say, 0.10 inch³/min., inch would require 70 lb/inch at 4,000 f.p.m. but only 30 lb/inch at 12,000 f.p.m. On the right graph, the plotting of Z_w (inch³/minute) versus grinding Horsepower yielded three linear relationships. The slope here has the units "horsepower used per cubic inch per minute" removal rate and is the specific power (P_{sp}). Increasing the wheelspeed reduced the P_{sp} from 6.7 at 4,000 f.p.m. to 4.0 HP/inch³ per minute as 12,000 f.p.m. Since the surface finish versus Z'_w performance also improved with increased wheelspeed, optimum use of CBN suggests using the highest wheelspeed that the grinding wheel marking, the grinding system, and the ANSI Safety Code will permit.

Fig. 5.7 shows the very beneficial effects of axial reciprocation upon finish. The upper graphs show that while reciprocation has no effect upon WRP with a resin-bond wheel (left graph), it will reduce the finish by about 10 microinches at all Z'_w removal rates. Essentially the same

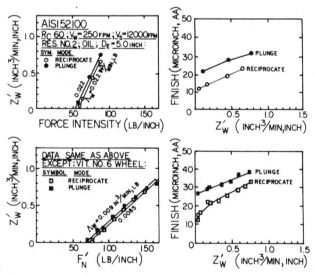

Fig. 5.7   Effect of axial reciprocation upon finish

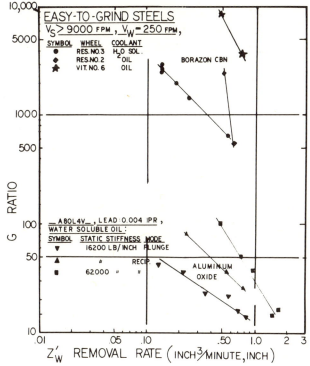

Fig. 5.8   Effect of metal removal rate on "G" ratio

results are shown in the lower graphs for a vitrified-bond wheel. Grinding of commercial-quality finishes with CBN is clearly achievable.

Fig. 5.8 illustrates that as the rough grinding metal removal rate, Z'_w, is increased, the G Ratio will decrease. That is true for both aluminum-oxide wheels at any system stiffness and CBN wheels using either water-soluble oil or sulfur-chlorinated oil. The latter fluid can, with the proper wheel, produce high G Ratios even at very high grinding rates.

B—#203 Inner Bearing Races

Size #203 ball bearings are the highest-volume bearing made in the world; hence much effort is made to reduce their internal-grinding cycle time. Understanding the grinding cycle is enhanced by using a portable recorder to measure cross-slide motion versus time. That gives a true picture of the grinding cycle, permitting comparisons to minimize cycle time. The bore size of a #203 race is 0.6693 inch.

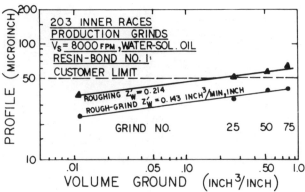

Fig. 5.9 Profile deterioration vs. grinding rates

Figure 5.9 shows the profile deterioration with continued usage when one is grinding at different rough grinding rates. At Z'_w of 0.143 inch3/ minute, inch at least 75 pieces whose profile never exceeded 50 micro-inches could be produced. However, when the removal rate was raised to 0.214 inch³/minute, inch the 50 microinch limit was exceeded after only 25 pieces (grinds).

It was decided to grind 203 races without dressing as long as the parts "looked good."

Figure 5.10 shows that when Proficorder traces of selected parts were made during the 350 part "no-dress" run, the hypothetical 50-microinch limit was exceeded probably after 36 to 40 pieces, and if that was to be a true limit, dressing should have been performed at least at that interval.

Today's Technology

CBN has been commercially available since 1970. It found fairly rapid acceptance in the high-speed tool maintenance (resharpening) opera-tions even though by 1985 that market is less than 20% converted to CBN. The acceptance of CBN in production grinding has been consid-erably slower, although it is expanding rapidly at this writing. The tech-nological knowledge in system and application engineering is advancing rapidly.

That technology is being pulled into the market by the larger tech-nically progressive companies that compete in world markets. The rate of expansion is slowed, however, because of several obstacles, which are now being overcome. Machine tool builders, with a few exceptions,

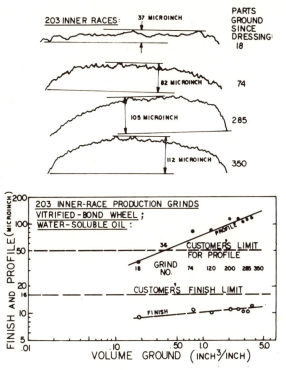

Fig. 5.10   Effect on wheel dressing of surface finish

have not generally taken an aggressive posture toward designing or modifying machines to accommodate CBN. Truing and dressing technology and systems are finally reaching commercial stages. A CBN wheel's acquisition cost, which can be up to 100 times that of a conventional wheel, is a general deterrent unless a total cost viewpoint is adopted. The bottom line is that CBN can reduce total grinding costs from 25% to 50% on a consistent basis.

As those obstacles are eliminated, CBN is being adopted in a wide variety of production grinding applications. They include internal, external, camshaft contour, centerless, surface, disc, creep feed, endmill feature, slot/groove, jig, honing, etc.

The fundamentals of most of those applications are discussed elsewhere in this text. Those elements of the system or the grinding process that are *different* and/or important for CBN and contribute to successful CBN application are offered below. Each application could be a chapter in itself. It is best to consult your grinding wheel supplier for competent application engineering assistance.

Internal

- machine must have "skip dress" capability
- wheel spindle truing compensation infeed should be .0001 inches radially or less
- tungsten carbide spindles or heavy metal quills control deflection during grinding
- vitrified is the bond of choice
- powered rotary diamond truing for production grinding

External

- 10% heavy-duty water-soluble fluid (if straight oil cannot be used)
- powered rotary diamond truing (generally) is necessary for wheels larger than 14 inches diameter and wider than $\frac{3}{4}$ inch
- truing compensation must be .00015 inches on radius or less
- truing traverse rate 15 to 20 inches per minute
- powered stick feed dressing is best for rapid and consistent "dressing"
- wheel "roundness" is very important—at least .0001, but 40 millionths inch (.000040) or less is preferred to prevent chatter
- automatic truing and dressing for high-production systems, such as camshaft contour grinding
- capability of changing the cycle to a more "balanced" cycle, i.e., lowering the "roughing" rate and raising the finishing rate to optimize the CBN wheel's work load.

Surface

- reciprocating table—resin is bond of choice, attention to truing/dressing, metal removal rates .5–.7 cubic inches per minute per inch of wheel width.
- Vertical spindle rotary table—attention to truing/dressing, resin or metal bond—if metal bond wheel should be slotted or of a "button" design, metal removal rates .1–.3 in ^{3}/min/inch; downfeed of .0015–.004 inches/minute.
- Creep feed—rigid machine designed for creep feed, adequate horsepower; coolant at least 10% heavy duty water soluble; high pressure 80–120 psi; high volume 40 to 60 gallons/minute, metal removal rate 1.0–2.0 in ^{3}/min/inch. resin bond for simple form; vitrified bond (crush roll or diamond roll truable) for more complex forms; plated for high-nickel materials (use straight oil) where finish requirement is 50 microinch or rougher.

• Disc—difficult to grind materials like M series high-speed steel or where tight tolerances are necessary, wet grind with 2 to 3% concentration of synthetic oil, coolant temperature 75–80°F, coolant flow $\frac{1}{2}$–$\frac{1}{3}$ normal, multiparticle nib or rotary truing (most efficient), truing infeed .0002 inches per pass maximum at 30–40 inches/minute traverse.

End Mill Features

Specially designed CNC machines, straight oil at high pressure 125–250 psi, kept filtered clean and at 95–105°F, low vibration, 20 millionths displacement or better, on machine dynamic wheel balance, true with rotary or multiparticle nib at .0002 inches radial infeed (max), dress with abrasive blast or stick.

Honing

Good part condition from previous operation, no heavy burrs, necks or gross out of round, hone speeds 250–300 surface feet per minute (90M/min); do not reverse stone direction to "free up the cutting" —this is not necessary with CBN, maximum CBN hone width $\frac{1}{8}$ to $\frac{3}{16}$ inch to assure clearing chips, conventional honing fluids are recommended with CBN.

What is Necessary to Expand CBN Usage Tomorrow?

An estimated 25% of the current conventional abrasive precision grinding market could be converted to CBN economically. Today's penetration is well below 5%. Continuing pressure for technical excellence, lower total costs, higher productivity, and improved part quality and materials will drive CBN's acceptance.

From a system/application engineering view the CBN product must be easy to use, which means machines designed to utilize CBN effectively with easy-to-use truing and dressing systems. Wheel suppliers must continue to build application knowledge and actively promote CBN capabilities. End users must think "total cost" instead of narrowly thinking "acquisition cost" or "abrasive cost/part" alone. CBN abrasive suppliers must work to get their costs down so prices of CBN wheels can be reduced. CBN wheels will always be many times more expensive than conventional wheels, but lower wheel costs will improve the economic justification process.

CBN is an ideal precision abrasive that can deliver very significant total cost and productivity improvements if properly applied. That is industry's challenge in the 1980s and 1990s—to properly apply that excellent cutting tool. That responsibility is shared by wheel suppliers, machine tool builders, and the end user. Many of the technical parameters for success are known. It is now necessary to establish easy-to-use commercial systems to utilize that technology.

REFERENCES

1.   E. Manson, Chien-Min Sung, "Design and Properties of Superabrasives," *Materials Science*, Spring 1980, 4–10.

2.   Wentorf, R.H. Jr., "Synthesis of Cubic Form of Boron Nitride," *J. Chen, Phys.*, 34 (1961): 809–12.

3.   Diamond Wheel Manufacturers Institute 712 Lakewood Center North, 14600 Detroit Avenue, Cleveland, Ohio, Publications: DWMI 101—Variables Affecting Diamond Wheels in Carbide Grinding; DWMI 105—Diamond Grinding of Carbides; DWMI 107—The Evaulation of metal Bonded Diamond Wheels for Grinding Non-Metalics.

4.   Subramanian, K., "Make the Best Use of CBN Grinding Wheels Through Proper Truing, Dressing, and Conditioning," *Superabrasives*, 1985 (DWMI/IDA/SME) April 1985.

5.   Lindsay, R.P., "Principles of Grinding with CBN Wheels" Diamond Wheel Manufacturers Institute Conference, Chicago, 1975.

CHAPTER 6

Grinding Chatter and Vibrations

K. Srinivasan
The Ohio State University

Introduction

Grinding is a widely used machining process in applications requiring high production rates and very good dimensional accuracy and surface finish. It is also one of the more complex metalworking processes and one of the less understood[1]. Consequently, successful use of grinding in practice is highly dependent on the level of expertise of the machinist and engineer. Grinding is also viewed as an unpredictable process because of the large number of variables involved and inadequate understanding of the relationships between those variables and the grinding process performance. That is particularly true of vibration in grinding operations, commonly referred to as grinding chatter.

Grinding chatter poses many of the same problems that chatter in other machining operations presents. Chatter results in undulations or roughness on the grinding wheel or workpiece surfaces and is highly undesirable. One form of grinding chatter, referred to as self-excited chatter, is usually eliminated or reduced by lowering metal removal rates. Also, grinding wheel surface unevenness resulting from chatter necessitates frequent wheel redressing. Thus, chatter results in a worsening of surface quality and lowers machining productivity. Those limitations are particularly severe, since grinding operations are used in applications involving high production quantities and stringent dimensional accuracy and surface finish requirements. It is also well recog-

119

nized that grinding chatter is a significant constraint on the selection of grinding conditions during process optimization or planning[1–5].

There are, however, a number of significant differences between grinding chatter and chatter in other machining operations, such as turning and milling. Those differences make it difficult to apply results related to chatter in turning and milling to grinding chatter. First, wheel-wear rates in grinding are significant when compared with workpiece material removal rates. Thus, wheelwear affects the geometry of the machining process, and hence grinding chatter, significantly. Second, the elasticity of the grinding wheel plays an important role relative to grinding chatter, since it affects the geometry of the contact zone between the grinding wheel and the workpiece. Local elastic deformations are of the same order of magnitude as the depth of cut in grinding but are much less significant in turning and milling. Third, since grinding wheels are composite structures in which the abrasive grains are randomly oriented in a matrix of bonding material, their geometry and mechanical properties are far more difficult to characterize than cutting tool geometry and mechanical properties in turning and milling. Finally, grinding processes are inherently more susceptible to chatter than turning and milling, and a large number of production grinding operations are dynamically unstable against self-excited chatter; that is, the vibration amplitudes grow with time, although they usually do so very slowly. The objectives of chatter analysis in grinding are therefore to help select grinding conditions or otherwise modify the grinding process to reduce chatter growth rates even if chatter cannot be eliminated.

Forced chatter, in contrast, refers to vibrations occurring in response to persistent disturbances, such as wheel or workpiece imbalance, wheel runout or offset and imbalance in other rotating components of the grinding machine. The amplitude of such vibration is relatively independent of the duration of the grinding operation, and remedies for reducing those amplitudes are described in a subsequent section. It will be seen then that many of the same remedies for reducing self-excited chatter amplitudes will reduce forced chatter amplitudes as well[6].

Even though a distinction can be made between self-excited and forced chatter in grinding, as indicated above, it is considerably more difficult, in practice, to label specific chatter problems clearly as being of 1 type or the other. Hahn[7] has offered guidelines for classifying grinding chatter patterns on workpieces in external and internal cylindrical plunge grinding and inferring underlying causes.[7] Surface imperfections resulting from chatter are often not measurable but are only visually observable. In other cases the surface imperfections can be measured from Talyrond charts and other measuring systems.

Improved productivity and workpiece surface finish would, therefore, result from better understanding of grinding chatter. Other expected benefits from the improved predictability and controllability of the grinding process include reduction in operator skill requirements and in setup time for batch manufacturing applications[1].

Types of Grinding Chatter

Grinding chatter and vibrations may be classified broadly into 2 classes:

1. Self-excited chatter or vibration and
2. Forced chatter or vibration.

Self-excited chatter refers to the vibrations resulting from the response of the grinding process to transient or short-lived disturbances. Those disturbances may be in the form of events, such as initial engaging of the wheel, and workpiece, or in the form of geometrical perturbations, such as initial undulations or unevenness on the wheel or workpiece surface. The vibrations occurring in response to the transient disturbance will decay with time if the grinding process is dynamically stable and increase with time if the process is unstable. The dynamic stability of the process is best assessed in terms of a dynamic model of the grinding process that accurately represents relationships between the grinding process geometry and machining forces. There is thus a closed-loop or feedback effect inherent in the grinding process as in other machining processes; that is, variations in the grinding process geometry affect the grinding forces and vice versa. Closed loop systems can become unstable depending on the dynamic relationships between variables within the loop. Fig. 6.1 is a highly simplified schematic representation of this closed-loop effect and is referred to as the chatter loop. The grinding forces and geometry referred to in the figure are best thought of as the summation of equilibrium values and deviations from the equilibrium values. Closed-loop systems such as the one in the figure are considered to be dynamically stable if the deviations from equilibrium, occurring in response to a transient disturbance, decay with time and they are considered to be unstable if the deviations increase in magnitude with time. Even though chatter in other machining processes such as turning may be examined within the same framework, there are significant differences between the grinding process and other machining processes as far as the details are concerned, as has already been noted. The dynamic modeling of different grinding processes and their relationships to self-excited chatter, formulated using chatter analysis results, are also described in these sections. For the present, it is important to recognize

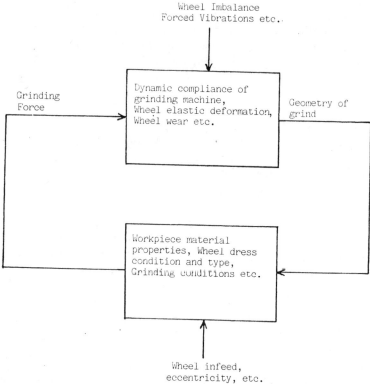

Fig. 6.1   Chatter loop in grinding

that the term self-excited chatter, as used here, refers to chatter occurring in response to transient disturbances and that its development depends significantly on grinding time.

Chatter patterns may be classified as straight, spiral or mottled[7]. If the chatter patterns on the workpiece are straight, the corresponding frequency of vibration can be determined from the spacing of the chatter marks and the workpiece surface speed. If that frequency is the same as the rotational frequency of the grinding wheel, wheel imbalance or wheel offset is the probable reason. If the frequency is twice the wheel rotational frequency, the probable reason is an asymmetric spindle, which has 2 principal directional modes of vibration or misalignment of spindle bearings, if the spindle has 3 or more bearings for support. If the vibrational frequency computed from the chatter marks is not so simply related to the wheel rotational frequency but does change proportionately when the wheelspeed is changed, the underlying reason could be vibration originating from other moving or rotating parts in the grinding machine. All of those cases represent instances of forced vibration. If the vibrational frequency is essentially independent of wheel-speed or workspeed but is close to a natural frequency of the wheelhead, workhead, or grinding machine structure, the underlying mechanism for the vibration is self-excited chatter. If the self-excited chatter occurs at low workspeeds and is accompanied by the development of lobes on the wheel surface that grow very slowly, the chatter is referred to as wheel regenerative chatter. If the self-excited chatter occurs at high workspeeds and is accompanied by the development of lobes on the work surface that grow rapidly, the chatter is referred to as workpiece regenerative chatter.

Spiral chatter patterns in plunge grinding operations are the result of vibration during wheel dressing. In traverse grinding operations, any of the mechanisms listed above could result in spiral chatter patterns. Mottled chatter patterns refer to random chatter patterns that do not fall on a straight line. If the chatter pattern repeats at the wheelhead frequency, the reason is most likely local hardness and stiffness variations in the grinding wheel.

The classification described above is useful if one chatter mechanism is dominant over others in a given situation. In other cases there may be a number of significant chatter mechanisms operational, leading to chatter patterns that are not easily classifiable according to the scheme described above.

Self-Excited Chatter in External and Internal Cylindrical Grinding

External and internal cylindrical grinding are the most widely used grinding processes and are more susceptible to self-excited chatter than other grinding processes, such as surface grinding. A model of the chatter loop in cylindrical grinding is described below, followed by a description of stability analysis techniques that help relate chatter behavior of the grinding process to process parameters. The dominant process parameters are then discussed at some length along with their effect on chatter.

Chatter Loop in Cylindrical Grinding

One of the most extensive models of the chatter loop in cylindrical grinding was proposed by Snoeys and Brown[8]. Fig. 6.2 describes the geometry of the cylindrical grinding process relevant for chatter. A normal cut in an external cylindrical grinding operation is represented, but

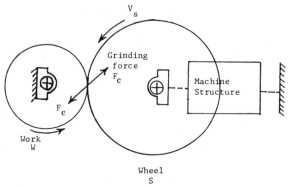

Fig. 6.2   Geometry of external cylindrical grinding

the results of the modeling and analysis can be applied also to internal grinding, as described later.

Infeed of the grinding wheel S into the workpiece W results in a corresponding grind force F_c. In response to the grinding force, workpiece stock is removed, the wheel wears, the workpiece and wheel deflect depending on the compliance of the grinding machine structure, and the contact zone between the wheel and workpiece is deformed because of wheel elasticity. The geometrical constraint that must be satisifed is that, at any instant of time, the total infeed u must equal the sum of the total wheel wear δ_s, total depth of stock removed δ_w, machine deflection y_m and the deflection y_k of the contact zone. Employing simple linear models for the phenomena listed above, Snoeys and Brown[8] derive the chatter loop shown in Fig. 6.3. Dynamic relationships within the chatter loop are represented by transfer functions, since the relationships are assumed to be linear and time invariant. Block diagram representation of linear dynamic system behavior is commonly used in linear dynamic system analysis[15].

The contact area stiffness K relates the contact zone deflection y_k and the grinding force F_c. The machine structural dynamic compliance is assumed to be represented by the transfer function $G_m(s)/K_m$. The contact zone has a finite length ℓ_c, which acts as a spatial filter and attenuates the effect of surface undulations on the wheel and workpiece. Short-wavelength undulations are attenuated more than long-wavelength undulations. The contact length filter transfer functions $Z_s(s)$ and $Z_w(s)$ are given later on. The instantaneous depth of stock removed $\Delta\delta_w$ and the instantaneous wheelwear depth $\Delta\delta_s$ are assumed to be proportional to the filtered grinding force.

The total wheelwear is given by adding instantaneous depth of wheel-

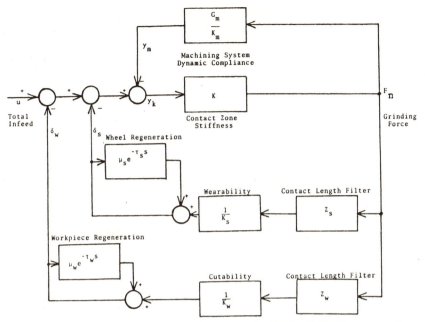

Fig. 6.3   Block diagram of chatter loop in cylindrical grinding[8]

wear $\Delta\delta_s$ to the wheelwear $\delta_s(t-\tau_s)$ during the previous period of wheel revolution. τ_s is the time period for 1 wheel revolution. Thus, wheel surface undulations created during the previous period of wheel revolution regenerate or affect the machining process during the current period of wheel revolution. The regenerative effect is a very important factor in determining the stability of the chatter loop. In plunge grinding applications where the grinding wheel feed direction is normal to the finished surface, the regeneration is complete. The overlap factor μ_s representing the overlap of the grinding wheel surface between 2 consecutive periods of wheel revolution is unity. In traverse grinding applications where the grinding wheel feed direction is parallel to the finished surface, the overlap factor μ_s is less than but very close to unity in practice. The workpiece regenerative effect is analogous to the wheel regenerative effect. The overlap factor μ_w, associated with workpiece regeneration is unity for plunge grinding applications and less than unity for transverse grinding applications. τ_w in Fig. 6.3 is the time period for 1 workpiece revolution.

The model of Snoeys and Brown[8] includes chatter loop models proposed by other researchers as special cases. Bartalucci and Lisini[9] consider the usual case in grinding applications where the workpiece

surface speed and rotational speed are much smaller than corresponding wheelspeeds, usually by 1 to 2 orders of magnitude. In such a case
workpiece surface undulations have short wavelengths especially if the
chatter frequency is high. For example, assuming wheel and workpiece
surface speeds of 30 m/sec and 0.3 m/sec respectively and a chatter frequency of 500 Hz, the wheel and workpiece surface undulations have
wavelengths of 60 cm and 0.6 mm respectively. The contact zone length
depends on the depth of cut and the wheel and workpiece sizes as well.
A typical value of the contact zone length[10] is 1-2 mm. The spatial filtering effect of the workpiece surface undulations is therefore considerable.
There is practically no filtering of the wheel surface undulations. That
condition is represented by letting the contact length filters $Z_s(s)$ and
$Z_w(s)$ in Fig.6.3 assume the values of unity and zero respectively. The
resulting chatter loop is shown in Fig. 6.4 and is said to represent the
case of wheel regenerative chatter. That occurs at low surface speeds
and high chatter frequencies, is characterized by very low chatter
growth rates, and is accompanied by the formation of surface undulations on the grinding wheel. Surface undulations on the workpiece are
not perceptible until wheel surface undulations have grown sufficiently
large. The grinding process can be continued for a few minutes before
there is a need for wheel redressing. Bartalucci and Lisini[9] and Inasaki
and Yonetsu[11] have proposed models for wheel regenerative chatter
that are very similar in form to Fig. 6.4. The differences between the
chatter loop models proposed by different researchers is primarily in the

Fig. 6.4   Chatter loop - wheel regenerative chatter[8]

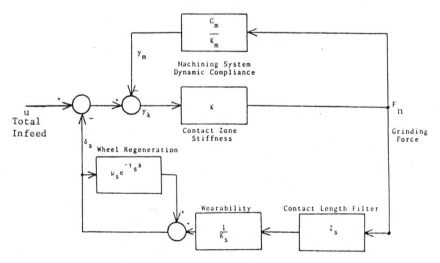

proposed relationships between grinding process parameters and the chatter loop model parameters K, K_s and K_w in Fig. 6.4.

The more general chatter loop in Fig. 6.3 represents the case of workpiece regenerative chatter. Both workpiece and wheel regeneration are effective. The contact length filter frequency responses are assumed by Snoeys and Brown[8] to be represented by

$$Z_w(j\omega) = \begin{cases} 1 & \dfrac{\omega}{2\pi} < \dfrac{V_w}{2\ell_c} \\[2mm] 0 & \dfrac{\omega}{2\pi} > \dfrac{V_w}{2\ell_c} \end{cases} \tag{6.1}$$

for the workpiece and

$$Z_s(j\omega) = \begin{cases} 1 & \dfrac{\omega}{2\pi} < \dfrac{V_s}{2\ell_c} \\[2mm] 0 & \dfrac{\omega}{2\pi} > \dfrac{V_s}{2\ell_c} \end{cases} \tag{6.2}$$

for the grinding wheel. In those equations,

$Z_w(s)$, $Z_s(s)$ = contact length filter transfer functions for the workpiece and grinding wheel respectively.

V_w, V_s = Surface speeds for the workpiece and grinding wheel respectively

ω = Frequency, rad/sec

and ℓ_c = Contact zone length.

At high workspeeds V_w and low chatter frequencies ω, therefore, workpiece regeneration is effective in addition to wheel regeneration. Workpiece regenerative chatter is characterized by high chatter amplitude growth rates and the formation of surface undulations on the workpiece surface even in the early stages of chatter. Only a few seconds of machining are possible in those cases before the workpiece surface undulations become clearly detectable. Inasaki and Yonetsu[11] have formulated a simpler model by ignoring the wheel regenerative effect. The reason for doing this is so that the wheelwear stiffness K_s, is usually a few orders of magnitude higher than the cutting stiffness K_w. Consequently, the contribution of wheelwear to workpiece regenerative chatter is small. Thompson[12] has also formulated a model for workpiece regenerative chatter. The elasticity of the contact zone is ignored in that model though it does take account of a small difference in the grinding geometry between climb cuts and normal cuts.

Chatter Loop Stability Analysis

The objectives of chatter loop stability analysis are to determine if the chatter loop in Figs. 6.3 or 6.4 is stable, for specified grinding conditions. An additional objective, for wheel regenerative chatter, is to determine chatter growth rates to enable either a determination of chatter-free grind time or evaluation of alternative grinding conditions to optimize chatter.

Since the chatter loop models assume constant parameters for given equilibrium conditions, the dynamic equations governing chatter behavior are linear difference-differential equations with constant coefficients. The stability of the chatter loop depends therefore on the location of the roots of the chatter loop characteristic equation in the complex s-plane, according to feedback control system theory.[15] The characteristic equation of the chatter loop for wheel regenerative chatter is

$$\frac{Z_s(s)}{K_j} \cdot \frac{1}{\mu_s e^{-\tau s^s} - 1} - \frac{1}{K} = \frac{G_m(s)}{K_m} \qquad (6.3)$$

and for workpiece regenerative chatter is

$$\frac{Z_w(s)}{K_w} \cdot \frac{1}{\mu_w e^{-\tau w^s} - 1} + \frac{Z_s(s)}{K_s} \cdot \frac{1}{\mu^s e^{-\tau s^s} - 1} - \frac{1}{K} = \frac{G_m(s)}{K_m} \qquad (6.4)$$

If all the characteristic equation roots lie in the left half of the complex s-plane, that is, have negative real parts, the closed loop system is stable against self-excited chatter. If 1 or more of the characteristic equation roots lie in the right half of the complex s-plane, that is, have positive real parts, the closed loop system is unstable against self-excited chatter.

One method of stability analysis of the chatter loop is to compute the roots of the characteristic Eq. 6.3 and 6.4 explicitly. The characteristic equations are transcendental equations because of the exponential terms and have an infinite number of roots. Standard iterative numerical procedures for solving for the roots of nonlinear algebraic equations can be used. The solution procedure should be capable of handling complex roots. Such solution procedures are available as part of many standard computer software subroutine packages, such as IBM's scientific subroutine package (SSP) and the IMSL software package.[17] The limitation of that method of stability analysis is that there is a significant amount of computation involved because of the infinite number of characteristic equation roots. The roots have to be computed 1 at a time, and different initial guesses for the root locations have to be provided for computing different roots. Also, as with any iterative scheme, convergence of the numerical method is not guaranteed unless the initial guess is close enough to the correct root location. For those reasons, explicit computation of the characteristic roots is rarely used to perform grinding chatter loop stability analysis in practice, though it is useful as a means for validating more convenient but approximate stability analysis techniques.[18-19]

Bartalucci and Lisini[9] have computed and plotted the characteristic equation roots for some cases of wheel regenerative chatter. Figure 6.5 shows 1 such case. The roots shown are the ones closest to the imaginary axis and hence critical for chatter loop stability. All other roots shown in the figure are in the right half of the complex s-plane. Some of the roots shown in the figure are in the right half of the complex s-plane, indicating that the chatter loop is unstable. The real part of the root farthest to the right, α_{max}, can be used as a quantitative measure of the self-excited chatter growth rate, and the corresponding imaginary part indicates the chatter frequency. Roughly speaking, the chatter may be expected to double at intervals of $0.69/\alpha_{max}$, quadruple at intervals of $1.39/\alpha_{max}$ and increase tenfold at intervals of $2.30/\alpha_{max}$. For the example in Fig. 6.5, α_{max} is 0.029 sec^{-1} and the chatter amplitude increases tenfold at intervals of 79 seconds. The chatter frequency is 399 Hz. Inasaki and Yonetsu[11] have considered a number of cases of wheel regenerative chatter and workpiece regenerative chatter and have computed closed loop characteristic root locations for those cases based on linear models of the chatter loop. α_{max} was computed to be in the range 0.001–0.002 sec^{-1} for the cases of wheel regenerative chatter considered and 0.1–1.5 sec^{-1} for the cases of workpiece regenerative chatter considered. Com-

Fig. 6.5   Chatter loop characteristic roots

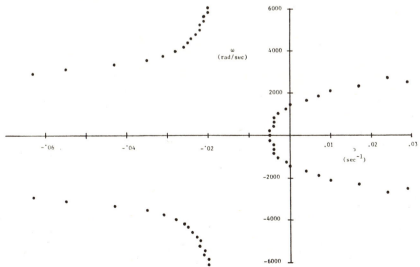

puted growth rates for workpiece regenerative chatter are therefore typically a few orders of magnitude higher than for wheel regenerative chatter. Experimental work by Inasaki and Yonetsu indicated that chatter amplitude in 1 case of workpiece regenerative chatter increased sevenfold in 20 seconds. On the other hand, chatter amplitude in 1 case of wheel regenerative chatter increased by a factor of about 3.5 in 16 minutes. Those experimental results are thus in good qualitative agreement with the chatter growth rates predicted from stability analysis.

An alternative technique of stability analysis involves a graphical procedure and requires that the chatter loop characteristic equation be written as

$$1 + G(s) = 0 \qquad (6.5)$$

The frequency response of $G(j\omega)$ is plotted, for the frequency ω varying from 0 to ∞, on a complex plane whose coordinates are the real and imaginary parts of $G(j\omega)$. That plot is called the Nyquist plot. If the plot does not encircle the $(-1, j0)$ point on the complex $G(j\omega)$ plane, and if the transfer function $G(s)$ has all of its poles in the left half of the complex s-plane, then the characteristic Eq. (6.5) has all of its roots in the left half of the complex s-plane and the chatter loop is stable. The term "poles" of $G(s)$ refers to values of the complex variable s for which $G(s)$ is infinitely large and hence corresponds to the singularities of the function $G(s)$. The stability result stated above is a simplified version of the Nyquist Stability Criterion commonly used in feedback control system theory.[15] Eq. (6.3) and (6.4) can be restated in the form (6.5) and the Nyquist Criterion applied as described above. However, such a stability analysis technique, though accurate, does not indicate readily and clearly the effects of individual chatter loop model parameter on system stability. Secondly, that method of stability analysis does not enable simple determination of chatter growth rates, which is practically important for wheel regenerative chatter. Hence, the straightforward application of the Nyquist Stability Criterion is of little use for chatter loop stability analysis.

The approximate, though more convenient and more practically useful, version of the Nyquist Stability Criterion was developed by Merritt[20] to investigate the stability of the chatter loop in turning. This stability analysis procedure was applied by Snoeys and Brown[8] to investigate the stability of the chatter loop in grinding. The characteristic Eq. (6.3) of the chatter loop for wheel regenerative chatter is written as

$$-\frac{K_m}{K^s} \cdot \frac{1}{-\mu_s e^{-\tau_s s}} - \frac{K_m}{K} = G_m(s) \qquad (6.6)$$

using the approximation that $Z_s(s)$ is unity for typical values of chatter frequencies and wheelspeeds. The Nyquist Criterion can be applied to that system to show[20] that the system is stable against self-excited chatter if there are no solutions of the following equation, that is, no intersections between plots of the left and right sides of the equation.

$$-\frac{K_m}{K_s} \cdot \frac{1}{1-\mu\,e^{-\mathcal{T}_s\omega}} - \frac{K_m}{K} = G_m(j\omega) \qquad (6.7)$$

That is only a sufficient condition for stability and not a necessary one. Therefore, the stability criteria obtained thus would be conservative. To simplify the stability analysis further, use is made of the fact that the plot of the left side of the equation lies to the left of the vertical line with the real part $(-(K_m/2K_s)-(K_m/K))$, for all values of τ_s, μ_s, and ω. Details of that proof are given by Merritt.[20] Therefore, if the plot of $G_m(j\omega)$ lies to the right of that vertical line, as shown in Fig. 6.6, the grinding operation is stable for all wheel speeds N_s.

$$\mathrm{Re}\,(G_m(j\omega)) \ge -\frac{K_m}{2K_s} - \frac{K_m}{K} \qquad (6.8)$$

Fig. 6.6   Chatter loop stability analysis

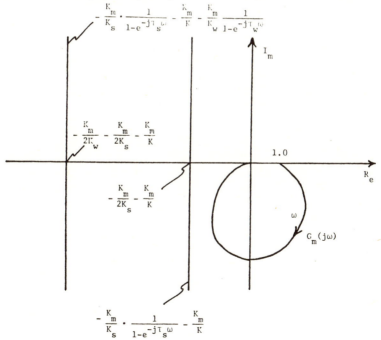

for stability, where Re refers to the real part. If the plot intersects the vertical line, the grinding operation may or may not be stable depending on the wheelspeed.

A stability chart for a grinding machine with known dynamic compliance characteristic $G_m(s)/K_m$ can be constructed for wheel regenerative chatter in plunge grinding ($\mu_s = 1$), in a manner similar to turning.[20] The abscissa is wheel speed N_s, and the ordinate is the inverse of the parameter ($(K_m/2K_s)+(K_m/K)$). Since K_s and K are proportional to grinding width in plunge grinding applications, the ordinate on the stability chart is proportional to grinding width and hence material removal rate. Once the stability chart is constructed for a machine, it is used as follows. The chatter loop model parameters are determined for a proposed grinding operation on the machine. The corresponding point on the stability chart is then located and indicates whether or not the proposed grinding operation is stable. Fig. 6.7 is an example of a stability chart for a grinding machine. It has a lobed appearance very similar to stability charts for turning operations. The grinding operations corresponding to points A, B, and C are stable, unstable, and marginally stable respectively. Details of the stability chart construction are given on page 154.

For the case of workpiece regenerative chatter, Snoeys and Brown[8] assume that chatter loop stability is equivalent to the absence of intersections between plots of the left and right sides of the equation below.

Fig. 6.7   Stability chart for grinding machine[6]

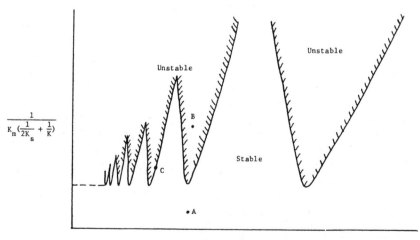

$$\frac{1}{K_m(\frac{1}{2K_s}+\frac{1}{K})}$$

Wheel Speed N_s

$$-\frac{K_m}{K_s} \cdot \frac{1}{1-\mu_s e^{-j\tau_s \omega}} - \frac{K_m}{K_w} \cdot \frac{1}{1-\mu_w e^{-j\tau_w \omega}} - \frac{K_m}{K} = G_m(j\omega) \qquad (6.9)$$

To simplify the analysis further, Snoeys and Brown[8] show that the plot of the left side of the equation lies to the left of a vertical line with the real part $(-(K_m/2K_s)-(K_m/2K_w)-(K_m/K))$ for all values of ω, μ_s, μ_w, τ_s, and τ_w. Therefore, if the plot of $G_m(j\omega)$ lies to the right of that vertical line, as shown in Fig. 6.6, the grinding operation is stable against workpiece regenerative chatter. Thus,

$$\text{Re}\,(G_m(j\omega)) \geq -\frac{K_m}{2K_s} - \frac{K_m}{2K_w} - \frac{K_m}{K} \qquad (6.10)$$

for stability. If the plot intersects the vertical line, the grinding operation may or may not be stable depending on wheel and workpiece speeds. Stability charts for grinding machines for regenerative chatter are not very practical, since they would be 3-dimensional because of the addition of workpiece speed N_w as a third coordinate.

The primary advantage of the stability analysis technique of Merritt[20] and Snoeys and Brown[8] is that it is simple to use and illustrates clearly the effect of different chatter loop parameters on stability. The relationships obtained thus are described in the following section. The disadvantage of the method is that it does not enable a simple determination of chatter growth rate in case of instability, which is of practical importance for wheel regenerative chatter.

A third technique of chatter loop stability analysis that enables the approximate determination of chatter growth or decay rates in addition to determining whether the chatter loop is unstable or stable has been developed recently. It utilizes the fact that, for many machining processes, the location of the roots of the chatter loop characteristic equation can be related simply to a function of frequency termed the regeneration spectrum. The regeneration spectrum has the important advantage that it is computed very easily, unlike the characteristic roots themselves. Consequently, the regeneration spectrum is a very effective tool in chatter loop stability analysis. The regeneration spectrum has been used effectively for the study of wheel regenerative chatter[18] and workpiece regenerative chatter.[19]

The regeneration spectrum is defined for time-delayed systems whose characteristic equation is of the form

$$P(s) + Q(s)e^{-sT} = 0 \qquad (6.11)$$

where P(s) and Q(s) are polynomials in s and T is the time delay. The regeneration spectrum is a plot of R(ω) versus the frequency ω.

$$R(\omega) = \left| \frac{Q(j\omega)}{P(j\omega)} \right| \tag{6.12}$$

The regeneration spectrum can be given a physical meaning for machining processes characterized by large values of the time delay T. In those cases the regeneration spectrum indicates the change, from 1 period of revolution to the next, of the spectral composition of transients in the system.

Under conditions stated by Srinivasan,[18] we can estimate features of the characteristic root distribution for time-delayed systems from the regeneration spectrum, with great savings in computation. The real parts of the characteristic roots are given approximately by Eq. (6.13)

$$\alpha_i = \text{Re } s_{i,-i} \simeq \frac{\ln (R(\omega_i))}{T} \tag{6.13}$$

where $s_{i,-i}$ are complex conjugate roots with their imaginary parts equal to ω_i in magnitude. In other words, the continuum of complex numbers $\ln[R(\omega)]/T \pm j\omega$ for ω varying from 0 to ∞ contains the important characteristic root locations of the system. In addition, it has been found that the spacing of the characteristic roots on the imaginary axis is approximately equal to $2\pi/T$ rad/sec. Therefore, the roots are closely spaced for large values of the time delay T. Under such conditions, the real part of the least stable characteristic root may be approximated by Eq. (6.14)

$$(\alpha_i)_{max} = \max_{i}\{\text{Re } s_{i,-i}\} \simeq \max_{0<\omega<\infty} \frac{\ln(R(\omega))}{T} \tag{6.14}$$

Using Eq. (6.14), we can readily determine the absolute and relative stability of the time-delayed system.

Eq. (6.13) may be viewed as an approximate version of the exact characteristic equation of system (6.11) stated as

$$\alpha_i = \text{Re } s_{i,-i} = \frac{\ln\left|\dfrac{Q(\alpha_i+j\omega_i)}{P(\alpha_i+j\omega_i)}\right|}{T} \tag{6.15}$$

If α_i is replaced by zero on the right-hand side, we obtain Eq. (6.13).

The system is unstable against chatter if and only if the regeneration spectrum has values exceeding unity for some frequencies ω. The growth of chatter amplitude is approximated by $e^{(\alpha_i)_{max}t}$. The frequency ω_{max} corresponding to the regeneration spectrum peak is a good approximation of the chatter frequency.

The characteristic equation for wheel regenerative chatter (6.3) can be rewritten as Eq. (6.16).

$$\left(\frac{1}{K} + \frac{Z_s(s)}{K_s} + \frac{G_m(s)}{K_m}\right) - \mu_s\left(\frac{1}{K} + \frac{G_m(s)}{K_m}\right)e^{-\tau_s s} = 0 \qquad (6.16)$$

That equation agrees in form with Eq. (6.11), and the regeneration spectrum for wheel regenerative chatter is defined as

$$R_s(\omega) = \left|\frac{\mu_s\left(\dfrac{1}{K} + \dfrac{G_m(j\omega)}{K_m}\right)}{\dfrac{1}{K} + \dfrac{Z_s(j\omega)}{K_s} + \dfrac{G_m(j\omega)}{K_m}}\right| \qquad (6.17)$$

Fig. 6.8 is a plot for the regeneration spectrum for the plunge grinding operation considered in Fig. 6.5. The peak value of the regeneration spectrum is 1.001, indicating that the grinding operation is unstable. The corresponding frequency of 417 Hz is an estimate of the chatter frequency. The real part of the least stable root of the chatter loop characteristic equation is estimated to be 0.029 sec^{-1} from Eq. (6.14). That estimated value compares very well with the exact value of 0.029 sec^{-1} obtained from the exact characteristic root locations. The estimated chatter growth rates would therefore be accurate. The chatter frequency estimated from the characteristic root locations is 399 Hz, reasonably close to that obtained from the regeneration spectrum. Since the characteristic

Fig. 6.8   Regeneration spectrum - wheel regenerative chatter[18]

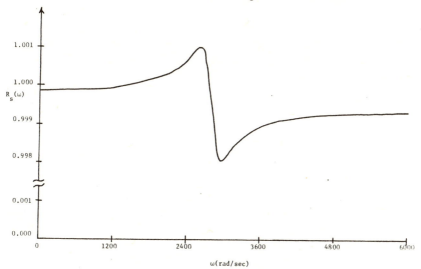

root spacing on the imaginary axis is approximately $2\pi/\tau$, rad/sec, the estimated chatter frequency and growth rate from the regeneration spectrum are more accurate for higher values of the time period of wheel revolution, τ_s. The computational effort required to obtain the regeneration spectrum is obviously much lower than that needed for the characteristic root locations.

The regeneration spectrum concept has been extended to the study of workpiece regenerative chatter in grinding by Srinivasan.[19] The corresponding characteristic Eq. (6.4) has 2 time delays, τ_s and τ_w. Two regeneration spectra are defined for that case. The workpiece regeneration spectrum $R_w(\omega)$ and the wheel regeneration spectrum $R_s(\omega)$ are defined below.

$$R_w(\omega) = \left| \frac{\mu_w \left[\dfrac{G_m(j\omega)}{K_m} + \dfrac{1}{K} \right] \left(\mu_s e^{-j\tau_s\omega} - 1 \right) - \mu_w \dfrac{Z_s(j\omega)}{K_s}}{\left[\dfrac{G_m(j\omega)}{K_m} + \dfrac{1}{K} + \dfrac{Z_w(j\omega)}{K_w} \right] \left(\mu_s e^{-j\tau_s\omega} - 1 \right) - \dfrac{Z_s(j\omega)}{K_s}} \right| \qquad (6.18)$$

$$R_s(\omega) = \left| \frac{\mu_s \left[\dfrac{G_m(j\omega)}{K_m} + \dfrac{1}{K} \right] \left(\mu_w e^{-j\tau_w\omega} - 1 \right) - \dfrac{\mu_s Z_w(j\omega)}{K_w}}{\left[\dfrac{G_m(j\omega)}{K_m} + \dfrac{1}{K} + \dfrac{Z_s(j\omega)}{K_s} \right] \left(\mu_w e^{-j\tau_w\omega} - 1 \right) - \dfrac{Z_w(j\omega)}{K_w}} \right| \qquad (6.19)$$

Eq. (6.19) for the wheel regeneration spectrum reduces to Eq. (6.17) if $Z_w(j\omega)$ is set to zero. Similarly, since the wheelwear stiffness K_s is a few orders of magnitude higher than the cutting stiffness coefficient K_w, Eq. (6.18) simplifies to Eq. (6.20) below.

$$R_w(\omega) = \left| \frac{\mu_w \left(\dfrac{G_m(j\omega)}{K_m} + \dfrac{1}{K} \right)}{\dfrac{G_m(j\omega)}{K_m} + \dfrac{1}{K} + \dfrac{Z_w(j\omega)}{K_w}} \right| \qquad (6.20)$$

Under conditions stated by Srinivasan[19] and shown to be true for workpiece regenerative chatter in grinding, we can estimate features of the characteristic root distribution for the chatter loop from the 2 regeneration spectra, with great savings in computation. The complex numbers $\ln[R_w(\omega)]/\tau_w \pm j\omega$ and $\ln[R_s(\omega)]/\tau_s \pm j\omega$, for ω varying from 0 to ∞ contain the important characteristic root locations of the system close to the imaginary axis. In addition, the spacing of 1 set of characteristic roots on the imaginary axis is approximately equal to $2\pi/\tau_w$ rad/sec and for the other set is equal to $2\pi/\tau_s$ rad/sec. The real part of the least stable characteristic root may be approximated as shown below.

$$\alpha_{max} \simeq \max_{0<\omega<\infty} \frac{\ln(R_w(\omega))}{\tau_w}, \text{ or } \frac{\ln(R_s(\omega))}{T} \qquad (6.21)$$

Therefore, the chatter growth or decay rate for the cylindrical grinding process can be determined readily from the regeneration spectra. The frequency ω_{max} corresponding to the peak value of the regeneration spectra is also a good approximation of the chatter frequency. Enhancements of the regeneration spectrum concept to improve the accuracy of Eq. (6.21) have also been described by Srinivasan.[19] In that same reference chatter growth rate and chatter frequency estimated from the workpiece and wheel regeneration spectra are shown to be in good agreement with values computed from the chatter loop characteristic roots.

It should be noted that the chatter growth rate computed using the regeneration spectrum method would be somewhat higher than the growth rate computed from the characteristic root locations. Also, the regeneration spectrum can be defined for chatter loop models more general than that of Snoeys and Brown.[8] The only requirement is that the model be linear and time-invariant.

Dominant Parameters in Self-Excited Chatter in Cylindrical Grinding

The chatter loop stability analysis techniques of the preceding section enable one to readily investigate the effect of individual chatter loop parameters on the stability of the grinding process. It can easily be shown[18,19] that the regeneration spectra for wheel and workpiece regenerative chatter, given by Eq. (6.17) and (6.20), can have values greater than unity indicating unstable grinding, only if the machine dynamic compliance has a negative real part for some frequencies ω. If instability does exist, larger values of the wheelwear stiffness K_s and of the cutting stiffness K_w result in lower chatter growth rates in wheel and workpiece regenerative chatter respectively. Larger values of the machine static stiffness K_m and lower values of the contact stiffness K result in lower chatter growth rates. Smaller values of the overlap factors μ_s and μ_w also result in lower chatter growth rates. Also, from Eq. (6.14) and (6.21), it is clear that for given peak values of the regeneration spectrum, high rotational speeds of the wheel and workpiece will result in higher chatter growth rates. Similar, though less general, conclusions regarding chatter growth rates can be deduced from the chatter stability analysis of Snoeys and Brown[8] and the Eq. (6.8) and (6.10).

The effect of the different grinding conditions on grinding chatter behavior can be studied by determining the resulting effect on the chatter loop parameters and using the results stated above to assess their effect on chatter loop stability. In order to do that, relationships between grinding conditions and chatter loop parameters are needed. Given the composite structure of grinding wheels and the multiple cutting edges involved, the number of variables needed to adequately describe the geometry of the grinding process is very large. The mechanics of the grinding process continue to be the subject of considerable research efforts. Therefore, analytical expression of chatter loop parameters are not available. Some empirical relationships developed for specified ranges of grinding conditions are available and are described on page 156. They are used below to formulate relationships between grinding conditions and chatter behavior. Experimental results illustrating such relationships are also described, where available.

Effect of Grinding Conditions

Wheel regenerative chatter is considered first, since chatter growth rates are low and permit continued grinding. The selection of grinding conditions to minimize chatter growth rates and maximize grinding times is thus a problem of great practical significance.[7] The exact relationship of grinding process parameters to chatter behavior would depend on the details of the grinding operation and the relationships between the process parameters and the chatter loop parameters. The example case of external cylindrical plunge grinding considered by Inasaki and Yonetsu[11] is used here in conjunction with the empirical relationships described on page 156 to derive the computed relationships given below. Eq. (6.29)–(6.32) are used for the cutting stiffness K_w, (6.37)–(6.38) for the wheelwear stiffness K_s (6.42) for the contact stiffness K, and (6.44) for the machine dynamic compliance. Specific values are given in Table 6—1, unless exceptions are noted. The chatter amplitude is assumed to grow or decay as $e^{(\alpha_i)_{max}t}$, where $(\alpha_i)_{max}$ is computed from the regeneration spectrum using Eq. (6.14). The results given below are obtained from Srinivasan.[18] Experimental results corresponding to those cases are not available.

Fig. 6.9 indicates the effect of the stock removal rate per unit width Z_w' on the chatter growth rate. The graph is approximately linear on a log-log scale. Eq. (6.38), (6.40), and (6.42) indicate that the wheelwear stiffness K_s decreases and the contact stiffness K increases as the stock removal rate Z_w' increases. Both of those trends result in higher chatter growth rates.

Increases in the workpiece surface speed V_w result in a lowering of chatter growth rates (Fig. 6.10, curve a). The workpiece surface speed is

Table 6—1   Details of typical grinding operation

Workpiece		Exceptions
D_w	40 mm	—
K_m	10 N/μm	—
ω_m	3142 rad/s	—
ζ_m	0.05	—
Rc	45	—
Material	hardened steel	—
Wheel		
Designation	2A60K6VLE	—
D_s	300 mm	Fig. 6-10, curve c
ℓ	0.10 mm/rev	—
c	0.013 mm	—
Vol	9.2 percent	—
d	0.38 mm	—
Feeds, Speeds		
W	10 mm	—
V_w	0.13 m/s	Fig. 6-10, curve a
V_s	28 m/s	Fig. 6-10, curve b
Z_w	16.1 mm³/min-mm	Fig. 6-9
Constants		
K_o	2680 N^0.25/mm	—

Fig. 6.9   Effect of stock removal rate on chatter growth rate[18]

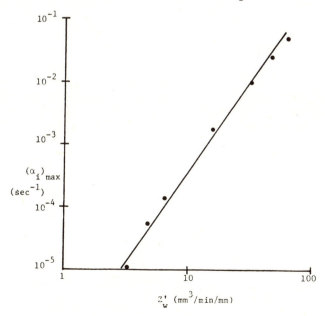

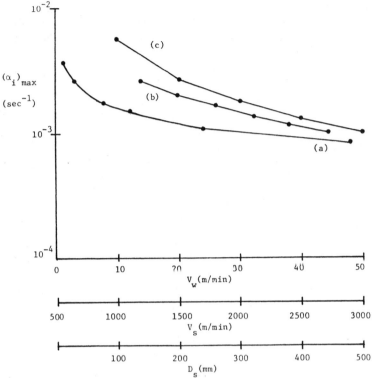

Fig. 6.10   Effect of grinding conditions on chatter growth rates[18]

assumed to be low enough in all cases for $Z_w(s)$, the contact length filter transfer function in Eq. (6.1), to be zero for the frequencies of interest, that is, wheel regenerative chatter predominates. An increase in V_w causes K_s to increase and K to decrease. Both those trends cause a lowering of the chatter growth rate.

Increases in the wheel surface speed V_s are seen to have a stabilizing effect, as shown in Fig. 6.10, curve b. K_s increases and K decreases as V_s increases, resulting in a lower peak value of the regeneration spectrum. However, the time delay τ_s decreases as V_s increases. According to Eq. (6.14), therefore, the chatter growth rate $(\alpha_i)_{max}$ should increase if the peak value of $R_s(\omega)$ remains the same. Apparently, the reduction in the peak value of $R_s(\omega)$ more than compensates for the reduction in τ_s, resulting in lower chatter growth rates.

Fig. 6.10, curve c, indicates that increasing the grinding wheel diameter D_s lowers chatter growth rates. An increase in D_s causes the equivalent diameter D_{eq} to increase. Eq. (6.38) and (6.40) indicate that K decreases and so does K_s, but to a much smaller extent. However, the time delay τ_s increases as D_s is increased, since the wheel surface speed V_s is constant. The stabilizing effect of a lower contact stiffness K and a higher τ_s dominate, resulting in lower chatter growth rates.

Relationships similar to those have been presented by Inasaki and Yonetsu[11] but were obtained following extensive computation. The transcendental characteristic equation of the chatter loop was solved numerically for each combination of grinding conditions. The regeneration spectrum method enables the same results to be obtained with a great saving in computation.

The effects of different grinding process parameters on the chatter growth rate exponent in external cylindrical grinding have been computed by Inasaki and Yonetsu[11] for similar grinding conditions and are noted in Fig. 6.11. The empirical relationships used for the chatter loop parameters are described in the reference noted above and are different from those used to derive the results in Figs. 6.9 and 6.10. As the depth of cut u_o increases, the stock removal rate per unit width Z_w' increases and the chatter growth rate exponent increases sharply, as shown by curve a in Fig. 6.11. That result is in qualitative agreement with Fig. 6.9. A larger grinding width w results in higher values of the contact stiffness K and the wheelwear stiffness K_s. The contact stiffness effect dominates and tends to increase the chatter growth rates, as shown by Fig. 6.11, curve b. An increase in workpiece surface speed V_w results in lower chatter growth rates for wheel regenerative chatter, as indicated by Fig. 6.11, curve c. That result is in agreement with Fig. 6.10, curve a. Inasaki and Yonetsu's analysis[11] indicates, however, that increased wheel speed increases chatter growth rates. That is different from the result in Fig. 6.10, curve b. The difference can be traced to the use of different models for the contact stiffness K. That brings up the point that the results in Figs. 6.9 to 6.11 are valid only if the empirical relationships on which they are based are valid.

Experimental results in support of many of the findings listed above are available. Inasaki and Yonestu[11] have found measured chatter growth rates in an external cylindrical grinding operation to be very low as predicted by their analysis and chatter frequencies in the range of 1.7–1.9 times the natural frequency of the single-machine vibrational mode. Their analysis indicates that the ratio should be 1.5. Bartalucci and Lisini[9] have found experimentally that chatter growth rates in wheel regenerative chatter increase with the mean normal grinding force intensity F_n'. Larger grinding force levels result in larger values of the contact stiffness K according to Eq. (6.42), which in turn leads to higher chatter growth rates. Snoeys and Brown[8] suggest that the chatter susceptibility of harder grinding wheels is due to the fact that for identical stock removal rates, preloads on harder wheels, and hence contact stiff-

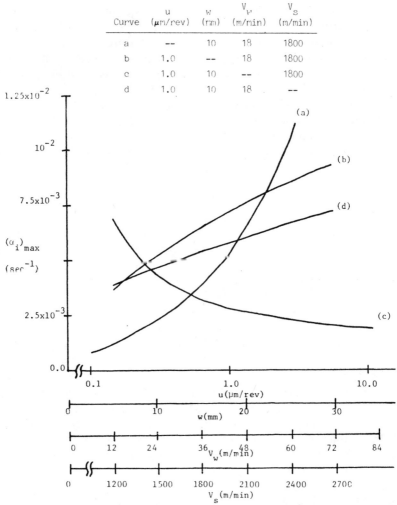

Curve	u (µm/rev)	w (mm)	V_w (m/min)	V_s (m/min)
a	--	10	18	1800
b	1.0	--	18	1800
c	1.0	10	--	1800
d	1.0	10	18	--

Fig. 6.11   Effect of grinding conditions on chatter growth rates in external cylindrical grinding[11]

nesses, are greater than for softer wheels. Lower contact stiffness is also the reason cited by Baylis and Stone[27] for the experimentally observed improvements in chatter behavior of specially designed grinding wheels. Bartalucci and Lisini[9] have also determined that chatter growth rates are very low and that chatter amplitudes vary exponentially with time in the initial stages of chatter development. That finding supports the characterization of chatter growth rates analytically by the chatter growth rate exponent in Figs. 6.9 to 6.11. In later stages of chatter development, the chatter amplitude decreases alternately[9] and the chatter frequencies drop.[8] That is ascribed to effects not included in the chatter loop, such as contact stiffness dependence on chatter amplitudes and metallic loading of the grinding wheel. Finally, experimental observations by Bartalucci and Lisini[9] and Snoeys and Brown[8] indicate that chatter amplitudes are modulated at the rotational frequency of the grinding wheel. That is a result of the fact that chatter signals consist of freqencies which are $1/\tau_s$ Hz apart. That finding in turn confirms the analytical results, such as those in Fig. 6.5, that indicate that characteristic roots are spaced $1/\tau_s$ Hz apart in frequency.

The discussion to that point has emphasized external cylindrical grinding but is applicable to internal grinding as well, with the use of the Eq. (6.33) instead of Eq. (6.32) for the equivalent diameter. There are a number of specific characteristics unique to internal grinding, however. As Hahn[32] has noted, the equivalent diameter in an internal grinding operation is higher for a given workpiece size and hence the contact length is greater. The filtering effect of the contact length is therefore greater and the regeneration of surface undulations is reduced, reducing chatter susceptibility. As Tonshoff[34] has noted, however, the same factors lead to higher radial wheelwear rates and potentially lower dimensional accuracy if that effect is not compensated for. Moreover, machine static stiffnesses are lower and natural frequencies are higher in internal grinding applications, as noted in the section on empirical Relationships. That results in higher susceptibility to chatter. Inasaki and Yonetsu[11] have computed chatter growth rates in wheel regenerative chatter for a range of grinding conditions very similar to that in Fig. 6.11. The results are noted in Fig. 6.12. The effect of changes in depth of cut u_o, width of cut w, and workpiece velocity V_w on chatter in internal grinding is similar to that in external cylindrical grinding, but the rates of chatter growth are considerably higher. That is a result of the lower static stiffness and higher natural frequency of the quill in internal grinding. Machine dynamic compliance cannot be altered very much in such applications either. Selection of grinding conditions to optimize chatter behavior in internal grinding is therefore of great significance.[18]

The effect of grinding conditions on workpiece regenerative chatter can be studied in a manner similar to that for wheel regenerative chatter. The emphasis, however, is on chatter avoidance rather than growth rate reduction because continued grinding is impractical in the presence of workpiece regenerative chatter. Workpiece regenerative chatter can be avoided most simply by reducing the workpiece surface speed adequately so that the contact length filtering effect becomes significant. Snoeys and Brown[8] have reported an example of external cylindrical grinding where reduction of the workpiece surface speed by a factor of 2.2 eliminated workpiece regenerative chatter and resulted in wheel regenerative chatter characterized by much lower chatter growth rates. Inasaki and Yonetsu[11] have indicated that, in a case of workpiece regenerative chatter in an experimental plunge cylindrical grinding operation, even though the grinding was stable at a depth of cut, u_o, it was unstable under transient sparkout conditions when the depth of the cut was lower. The chatter loop analysis presented by the authors indicates that such a behavior is possible if the cutting stiffness decreased with increase in the depth of cut, as in Eq. (6.36). That effect would counteract the destabilizing effect of higher contact stiffness at higher depths of cut. For high enough workpiece surface speeds, there would then be a range of depths of cut for which the grinding process is unstable. The process is stable for depths of cut higher and lower than that range. Inasaki and Yonetsu[11] recommend lowering the workpiece surface speed to avoid that type of behavior. Experimental work by Inasaki and Yonetsu also indicates high rates of chatter growth as indicated by analysis and a chatter frequency of 1.6 times the natural frequency of the single vibrational mode of the machining system. Analytical work for very similar grinding conditions indicated a chatter frequency of 1.5 times the natural frequency.

Effect of Machine Structure Redesign. The chatter loop stability analysis techniques of the previous section can be used to evaluate the effectiveness of proposed changes in the machine dynamic compliance on the chatter behavior. Changes in the machine structure that reduce the magnitude of the most negative real part of the machine dynamic compliance $G_m(j\omega)/K_m$ would improve chatter behavior of the machine. Brown[6] has confirmed that experimentally for a number of grinding machines. Some of the guidelines developed by Brown are given here. For machine tools with multiple vibrational modes, as in Eq. (6.45), weak high frequency modes can be used to compensate for low frequency modes. It is assumed here that the chatter frequency corresponds to the low-frequency mode and is lower than the natural frequency of the high-frequency mode. Therefore, the weak high-frequency mode contributes a positive real part at the chatter frequency and hence reduces the magnitude of the negative real part of $G_m(j\omega)/K_m$ at the chatter frequency. That in turn should reduce chatter growth rates. The weak high-frequency modes would be intentionally designed and associated most probably with the vibration of workpiece support or the wheel spindle. In fact, Brown[6] has cited instances where the elimination or stiffening of such high-frequency modes by grinding machine manufacturers has resulted in worsening of the chatter behavior. Therefore, contrary to the widely held belief, stiffening of all modes of structural vibration is not necessarily the best way to improve chatter behavior. Secondly, increase of the structure damping, by adding squeeze film or viscoelastic dampers, was shown by Brown to be effective in reducing chatter susceptibility of grinding machines That can be related to the fact that the negative real part of $G_m(j\omega)/K_m$ is reduced in magnitude by the addition of damping.

Evaluation of proposed machine structure redesign based on the real part of the dynamic compliance $G_m(j\omega)/K_m$ is appropriate if no information is available on the other chatter loop parameters. However, if improving the chatter behavior of a specific grinding process is the objective, a more accurate evaluation of proposed machine structure redesign is possible by using the regeneration spectrum. Srinivasan[18,19] has shown that the regeneration spectrum is related more closely than the real part of $G_m(j\omega)/K_m$ to chatter growth rates in wheel and workpiece regenerative chatter. Consequently, changes in the machine structure that would reduce the peak value of the regeneration spectrum would improve the chatter behavior of the process. The guidelines for machine structure redesign obtained thus by Srinivasan[18,19] are the same as those of Brown,[6] but the quantitative conclusions based on the regeneration spectrum are more accurate. Of course, all of the chatter loop parameters are needed to compute the regeneration spectrum. Hence that approach can be used only if the grinding conditions and the corresponding chatter loop parameters are known.

Self-Excited Chatter in Surface Grinding

Chatter in surface grinding has received considerably less attention in research literature than chatter in cylindrical grinding, since the latter operation is more widely used in practice. Since workpiece traversing speeds are usually lower in surface grinding than in cylindrical grinding, contact length filtering of the workpiece surface undulations is pronounced and results in negligible regeneration of workpiece surface undulations. Therefore, the reversal in the direction of workpiece motion and the consequent reversal in time of the regeneration of workpiece surface undulations is of little practical significance. Wheel regenerative chatter predominates and the chatter loop is modeled well by Fig. 6.4. The only difference is that in surface grinding the grinding wheel may be out of contact with the workpiece during table reversal at the end of the grinding strokes. That interruption in grinding could be represented by letting the contact zone stiffness K and the grinding force in Fig. 6.4 go to zero for the duration of the grinding wheel overrun. The interruption in grinding destroys the periodicity of the wheel regeneration since the time of wheel overrun is not necessarily an integral multiple of the period of wheel revolution. Chatter growth rates are therefore lower in surface grinding applications than in cylindrical grinding applications under similar conditions[8].

A mathematical model of the chatter loop in surface grinding, comparable in completeness to that for cylindrical grinding, is not available. Thompson[35,36] has formulated a model for chatter in surface grinding and presented some experimental results. However, Thompson's model ignores the elasticity of the grinding wheel that has been shown to be essential for proper description of the chatter loop in cylindrical grinding. It has also been noted by a number of researchers[37] that interrupted grinding, such as in surface grinding, brings into play some other thermal effects, such as the cooling of the wheel and workpiece during the grinding wheel overrun. That in turn results in lower grinding forces the longer the overrun time in relation to the grinding time. An accurate quantitative treatment of chatter in surface grinding would need to include that phenomenon in the chatter loop model.

The dominant parameters in self-excited chatter in surface grinding and their effects are the same as those for wheel regenerative chatter in cylindrical grinding and have been stated at great length in the previous section. Inasaki and Yonestu[11] have computed chatter growth rates as a function of depth of cut, width of cut and workpiece speed for conditions typical of surface grinding and have derived curves similar to corresponding curves for external and internal cylindrical grinding (Fig. 6.11 and 6.12). The expressions for the chatter loop parameters in terms of the grinding conditions in the section on Empirical relationships (page 157) can be used for surface grinding as well, if the wheel diameter D_s is used for the equivalent diameter D_{eq}. As has been noted in that section the static stiffness K_m of surface grinders is usually lower than that of cylindrical grinders. However, their overall susceptibility to chatter is less than those of cylindrical grinders because of the interrupted nature of the grinding.

Self-Excited Chatter in Centerless Grinding

Centerless grinding is susceptible to self-excited chatter, as are external and internal cylindrical grinding. A model of the chatter loop is described below, followed by results obtained using stability analysis techniques. The dominant process parameters governing chatter behavior and the rounding effect are then discussed briefly.

Curve	u (μm/rev)	w (mm)	V_w (m/min)	V_s (m/min)
a	--	10	30	1800
b	0.5	--	30	1800
c	0.5	10	--	1800

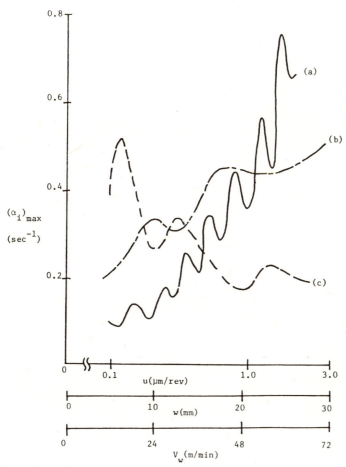

Fig. 6.12   Effect of grinding conditions on chatter growth rates in internal grinding[11]

Chatter Loop in Centerless Grinding

Miyashita et al,[5, 34-40] have investigated chatter in centerless grinding, theoretically and experimentally. Fig. 6.13 describes the geometry of centerless grinding relevant for chatter. The grinding wheel rotational

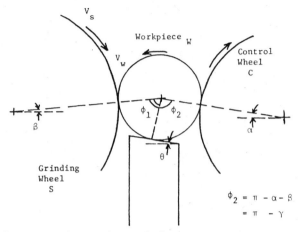

Fig. 6.13   Geometry of centerless grinding

speed is N_s, and the control wheel rotational speed is N_c. The workpiece rotates with a surface speed V_w almost equal to that of the control wheel. The workplate supporting the workpiece has its supporting face at an angle θ. α, β, γ, $\emptyset_1$ and $\emptyset_2$ are angles describing the relative oientation of the grinding and control wheels and the workpiece and are indicated on the figure. The geometry of the grind obviously depends on the total grinding wheel infeed, wear of the grinding and control wheels, the deflections of the machine structure, and the wheel deformations caused by the wheel elasticities. However, the cut geometry is also influenced by the contact between the workpiece and the workplate and the control wheel. Since previously machined portions of the workpiece surface are involved in those contacting areas, they give rise to the 2 additional regenerative terms in the block labeled as the geometrical rounding effect function in Fig. 6.14. The finite contact lengths at the grinding and control wheels have a spatial filtering effect on the workpiece surface undulations as in cylindrical grinding and have the transfer functions $Z_{ws}(s)$ and $Z_{wc}(s)$ respectively. The instantaneous depth of cut is also affected by the regenerative feedback of workpiece surface unevenness from the previous period of workpiece revolution. The cutting stiffness K_w determines the grinding force corresponding to the depth of cut. The elasticities of the grinding and control wheels result in local deformations at the contacting surfaces and affect the cut as do machine structural deflections. Fig. 6.14 is a block diagram representing those effects and has been proposed by Miyashita et al[40]. The wear of the grinding and control wheels has not been taken into account here though it has been mentioned by Miyashita et al. in earlier work[38, 39].

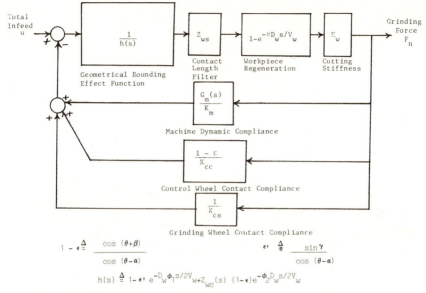

Fig. 6.14   Block diagram of chatter loop in centerless grinding

Therefore the chatter loop described here is for a case of workpiece regenerative chatter.

Chatter Loop Stability Analysis

The characteristic equation of the chatter loop in Fig. 6.14 is given by

$$\frac{1-\epsilon'e^{-\tau_1 s} + Z_{wc}(s)(1-\epsilon)e^{-\tau_2 s}}{K_w \, Z_{ws} \, (s) \, (e^{-\tau_w s}-1)} = \frac{G_m(s)}{K_m} + \frac{1-\epsilon}{K_{cc}} + \frac{1}{K_{cs}} \qquad (6.22)$$

The roots of the characteristic equation can be computed numerically, for specified values of the chatter loop parameters. If all of the root have negative real parts, the chatter loop is stable. If any of the roots have positive real parts, the chatter loop is unstable. The real part of the root with the largest real part, α_{max}, is a measure of the chatter growth rate and is called the chatter growth rate exponent. Numerical solution of the roots of the transcendental Eq. (6.22) is a tedious task, since it has 3 time delays τ_1, τ_2, and τ_w and 3 corresponding exponential terms.

Miyashita et al.[5] have used a graphical procedure for solving an approximate version of Eq. (6.22), obtained by making the simplifying assumptions

$$\epsilon = 0 = \epsilon'$$

$$Z_{wc}(s) = 1 = Z_{ws}(s) \tag{6.23}$$

and $G_m(s) = G_m(\alpha + i\omega) = G_m(i\omega)$ for small α

The last of the assumptions is justified by the fact that typically the chatter growth rate exponent α_{max} is very low. The approximate characteristic equation is

$$\frac{1+e^{-\tau_2 s}}{e^{-\tau_w s}-1} = K_w \left(\frac{G_m(i\omega)}{K_m} + \frac{1}{K_{cc}} + \frac{1}{K_{cs}} \right) \tag{6.24}$$

The right-hand side of the equation is graphed for frequencies ω varying from 0 to ∞. The left-hand side is a function of both α and ω. Miyashita et al.[40] use a graphical procedure for determining α and ω for which Eq. (6.24) is satisfied. It is noted that most centerless grinding operations are unstable against self-excited chatter. Typical experimental chatter growth rate exponents α_{max} reported by Miyashita et al.[5,39] are in the range of 0.03-0.1 sec^{-1}, resulting in doubling of chatter amplitudes at intervals 7–23 seconds. Depending on the grinding cycle time, therefore, grinding is practical even though the chatter loop is unstable.

The results of the stability analysis for a specific machine and grinding operation of interest are noted in the form of a "Rounding Effect Diagram." The coordinates on the figure are parameters depending on the chatter frequency, workpiece rotational speed, and the center-height angle γ. Contours corresponding to constant chatter growth rates are plotted on the figure. The chart is designed to help in the selection of the workpiece speed and the angle γ to minimize chatter growth rates, other grinding conditions being fixed[5].

The approximate stability analysis technique ignores the geometrical rounding effect in centerless grinding, represented by the geometrical rounding effect function in Fig. 6.14. Frequencies ω for which that function has very high gains correspond to workpiece surface undulations that are not ground out by continued grinding. Frequencies ω for which that function has low gains correspond to workpiece surface undulations where the geometrical rounding are most effective. Miyashita et al.[5,38-40] consider that effect independently of the remainder of the chatter loop and conclude that the following additional criteria must be met for effective geometrical rounding:

1. γ should be in the range of 6°–8°.
2. The workplate angle θ must be such that the ratio $\emptyset_1/\gamma$ is close to an odd integer.

Dominant Parameters in Self-Excited Chatter in Centerless Grinding

The chatter loop stability analysis techniques described in the preceding section enable one to draw conclusions regarding the effects of different chatter loop parameters on chatter behavior. Those conclusions are very similar to those for cylindrical grinding. Unstable grinding can occur only if the machine dynamic compliance $G_m(j\omega)$ has a negative real part for some frequencies ω. Larger values of the machine static stiffness K_m and lower values of the wheel contact stiffnesses K_{cc} and K_{cs} result in lower chatter growth rates. Lower wheel contact stiffnesses also result in more pronounced filtering effect of the workpiece surface undulations and lower chatter growth rates. That effect is not indicated by the simplified stability analysis above since Eq. (6.23) eliminates the filtering effect. The stabilizing effect of control wheel flexibility has been noted experimentally by Miyashita et al[5]. The significance of wheel elastic deformations compared with machine structural deflections has also been noted in the same reference.

The effect of grinding conditions on chatter behavior in centerless grinding has not been studied to the same extent as cylindrical grinding. However, the empirical relationships for the chatter loop parameters can be used in conjunction with the analysis above to assess those effects. Many of the relationships described for cylindrical grinding can be seen to be valid for centerless grinding as well. The geometrical rounding effect is unique to centerless grinding, however, and leads to the additional conditions noted above on the center-height angle γ and the workplate angle θ, for effective geometrical rounding.

Forced Vibration in Grinding

Forced vibration refers to vibrations occurring in response to persistent disturbing influences determined by factors external to the chatter loop. Fig. 6.15 shows a block diagram of the chatter loop in cylindrical grinding and indicates how some of those disturbing inputs affect the grinding process.

One of the remedies for reducing the severity of the forced vibration problem is to identify the sources of the disturbance and reduce the level of the disturbance. If the disturbance is in the form of vibrations transmitted through the floor to the machine structure, modification of the machine foundation using well-established techniques[41] would reduce the forced vibrations. If wheel imbalance caused by differences in wheel deflections or velocities during dressing and grinding is the problem,

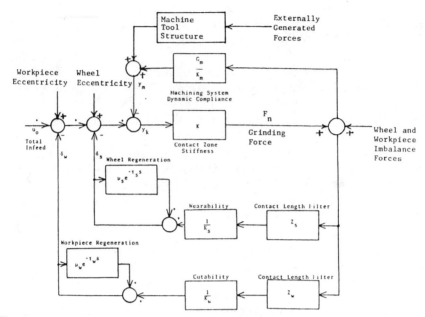

Fig. 6.15   Forced vibrations in cylindrical grinding

wheel dressing during grinding could reduce forced vibrations[37]. Similarly, minimizing spindle bearing clearances and misalignment would reduce forced vibrations caused by spindle imbalance.

It is also important to recognize that the surface finish imperfections resulting from forced vibration can be improved by improving the stability of the chatter loop. That is, many of the measures that improve the self-excited chatter behavior of the grinding operation would also improve the forced vibration response[6]. For example, stiffening of the machine tool structure and reduction of its dynamic compliance would improve the forced vibration response in general.

Finally, small changes in grinding conditions can sometimes result in large changes in the forced vibration response of the system. For instance, Brown[6] has reported that small changes in the workpiece speed in external cylindrical grinding resulted in significant improvements in forced vibration levels. That type of behavior is expected if the frequency of the disturbing signal originally coincides with a peak in the amplitude frequency response of the closed chatter loop. Because of the time delays in the chatter loop, there would be a number of such sharp peaks spaced at frequency intervals of $1/\tau_s$ and $1/\tau_w$ Hz respectively for wheel and workpiece regenerative chatter. Consequently, small changes in

workpiece speed could shift the peaks and reduce the res
chatter loop at the forcing frequency.

Techniques for Chatter Reduction

Techniques for chatter reduction can involve selection of appropriate grinding conditions, machine structure modification, and/or the use of special control strategies. Since continued grinding is feasible for many grinding operations despite growing chatter amplitudes and since there are other constraints relating to surface integrity, grinding temperature and machine limitations that cannot be violated, chatter behavior is only 1 among many constraints. Maris, Snoeys and Peters[3] and Hahn[2] have described procedures for optimizing grinding process design while explicity taking into account chatter behavior, as indicated by the stability analysis techniques described above. Such an approach ensures that all of the relevant considerations in grinding process design are incorporated explicitly. However, in many cases enough information may not be available to carry out a complete chatter loop stability analysis. In such situations recourse may be had to simpler guidelines of the type indicated by Maris, Snoeys and Peters[3] and Hahn[7].

Wheel regenerative chatter growth rates can be reduced by frequent wheel dressing, reducing feedrate and/or width of cut, use of a larger and/or softer wheel, and use of grinding fluids to reduce wheelwear. If high stock removal rates are planned, higher workpiece speeds can be used without encountering workpiece regenerative chatter. Workpiece regenerative chatter can also be avoided by reducing workpiece speed, increasing feedrate, and using a larger and/or softer wheel. Also, as Brown[6] has indicated, stiffening of weak low-frequency modes of vibration of the machine structure and the addition of damping will reduce the chatter susceptibility of machine tool structures. Weak high-frequency modes of vibration may actually be helping reduce chatter susceptibility of the machine tool structure. Increased damping in those modes would help improve chatter behavior further.

Other special control strategies can be used to reduce chatter growth rates. Variation of the grinding wheelspeed in wheel regenerative chatter would interfere with the regeneration mechanism and could result in lower chatter growth rates. Bartalucci and Lisini[9] have shown experimentally that wheel regenerative chatter amplitudes can be reduced by a factor of 2–10 by using a 10% wheelspeed variation. Cegrell[42] has reported experimental results that show a five-fold reduction of chatter amplitudes using a 10% wheelspeed variation. Pahlitzsch and Cuntze[43] have reported improved chatter behavior using an additional force for excitation of the workpiece. That force is required to have the same frequency as the chatter vibration and a fixed amplitude and phase relationship to the grinding force. Other control strategies have been listed by Snoeys and Brown[8].

Conclusions

It should be clear that the phenomenon of grinding chatter is a complex one. However, available chatter analysis techniques do provide conclusions and guidelines whose validity is borne out by practical results. Experimental techniques for obtaining information on the chatter loop parameters have not been described here but are described in the references cited in the chapter. Careful experimentation is necessary to obtain the data needed for a complete chatter analysis. Further improvements in our understanding of chatter behavior can be expected as our understanding of the mechanics of grinding improves, along with development of better techniques for characterizing grinding wheel composition and properties and appropriate modification of chatter analysis techniques.

Construction of Stability Chart for Wheel Regenerative Chatter

Stability charts for grinding machines have the same potential uses that similar charts for lathes and milling machines have. The procedure for constructing stability charts for wheel regenerative chatter in plunge grinding is also similar and is described below.

For a given plot of $G_m(j\omega)$, the upper limit for $((K_m/2K_s)+(K_m/K))$ stable grinding at all wheel speeds is determined by setting

$$\frac{K_m}{2K_s} + \frac{K_m}{K} = Re_{min}(G_m(j\omega)) \tag{6.25}$$

where the subscript "min" refers to the minimum of $Re(G_m(J\omega))$, evaluated over all positive ω. For all values of $((K_m/2K_s)+(K_m/K))$ lower than the limit in equation (6.25), the grinding operation is stable at some wheelspeeds and unstable at others. To determine the ranges of wheelspeeds for stable and unstable grinding, the following procedure should be followed:

(1) Choose a value of $((K_m/2K_s)+(K_m/K))$ lower than the limit in equa-

tion (6.25) and draw a vertical line, with - $((K_m/2K_s)+(K_m/K))$ as the real value, on the same figure as the $G_m(j\omega)$ plot. The 2 plots will intersect at a few points. Label the corresponding values of ω, from the $G_m(j\omega)$ plot, as ω_i, i = 1,2, etc. (Fig. 6-6).

(2) The frequencies ω_i represent chatter frequencies that would result if the wheelspeed is such that Eq. (6.7) is exactly satisfied. The accompanying grinding force variations would result in wheel surface undulations. Now,

$$\omega_i \tau_{si} = \frac{\omega_i}{N_{si}} = 2\pi\ (n_s + \nu_{si}) \qquad (6.26)$$

Here n_s is a positive integer and ν_{si} is a fraction between zero and 1. They represent the integral and fractional number of lobes that would form on the wheel periphery at the chatter frequency ω_i. From equation 6.26 we get

$$\frac{1}{1 - e^{-j\tau_{si}\omega_i}} = \frac{1}{1 - e^{-j\,2\pi\nu_{si}}} \qquad (6.27)$$

At the points of intersection obtained in step (i), therefore,

$$-\frac{K_m}{K_s} \cdot \frac{1}{1 - e^{-j2\pi\nu_{si}}} - \frac{K_m}{K} = G_m(j\omega_i) \qquad (6.28)$$

We can solve for ν_{si} since it is the only unknown for each point of intersection.

(3) Assume n_s is zero in equation (6.26). Using computed values of ω_i from step (1) and the corresponding ν_{si} from step (2), compute τ_{si} and N_{si} from (6.26). The N_{si} represents wheelspeeds for which the grinding process is marginally stable for the process parameter value $((K_m/2K_s)+:K_m/K))$ assumed in step (1). At other values of the wheelspeed, the grinding process is either stable or unstable depending on the wheelspeed. Arrange the computed wheelspeeds N_{si} in descending order. Let $(N_s)_{max}$ be the value of the highest N_{si}. For wheelspeeds higher than $(N_s)_{max}$, the grinding process is stable. For wheelspeeds between $(N_s)_{max}$ and the lower value of N_{si}, the grinding process is unstable. Regions of stable and unstable operation alternate with one another and the computed N_{si} indicates the boundaries between adjacent regions of stable and unstable operation.

(4) Repeat step (3) assuming successively higher integer values of n_s. For each value of n_s, determine N_{si} and assess the stability of the grinding operation for wheelspeeds between the computed N_{si}, using the procedure in step (3). As n_s is increased, the computed regions of unstable operation overlap with those computed for lower values of n_s,

indicating that higher values of n_s need not be considered. A proposed grinding operation is unstable if it lies within the unstable range of wheelspeeds corresponding to any value of n_s. If it does not, the grinding operation is stable.

(5) Steps (1) – (4) are repeated for lower values of $((K_m/2K_s)+: K_m/K))$. Since the wheelwear stiffness K_s and the contact zone stiffness K are proportional to the width of cut in plunge grinding applications, lower values of $((K_m/2K_s)+(K_m/K))$ represent higher values of widths of cut and hence higher rates of material removal. We should therefore expect that the wheelspeed ranges for unstable operation would widen and the wheelspeed ranges for stable operation would narrow as $((K_m/2K_s+K_m/K))$ is increased. For a low enough value of that parameter, the grinding operation would be unstable for all wheelspeeds of practical importance. Values of the parameter lower than that limit need not, therefore, be attempted.

(6) The stability chart for the grinding machine is constructed using wheel speed N_s as the abscissa and the inverse of $((K_m/2K_s)+(K_m/K))$ as the ordinate. It will have a lobed appearance similar to stability charts for turning operations.

Empirical Relationships for Chatter Loop Parameters

Empirical relationships for the chatter loop parameters K_w, K_s, K, and the contact length ℓ_c have been determined by a number of researchers for specified grinding conditions and are described below.

Cutting stiffness—Hahn and Lindsay[21-22] have related the cutting stiffness to the metal removal parameter Λ_w by the following equation.

$$K_w = \frac{V_w w}{\Lambda_w} \qquad (6.29)$$

where:

K_w = Cutting stiffness, N/m
V_w = Workpiece surface speed, m/min
w = Width of cut, mm

and Λ_w = Metal removal parameter, mm³/min-N

The metal removal parameter is related to grinding conditions and details of the wheel dressing preceding the grinding operations. Hahn and Lindsay[22] have formulated the following relationship, assuming single-point diamond dressing, for easy-to-grind workpiece materials.

Easy-to-grind materials are defined as materials that form chips easily. Cutting takes place at low values of the grinding force for such materials. In contrast, difficult-to-grind materials exhibit large rubbing and ploughing zones and require large values of the grinding force to initiate cutting. The relationship below for easy-to-grind materials was seen to be accurate to within 20% when evaluated over data from 400 grinds corresponding to 65 different values of Λ_w.

$$\Lambda_w = \frac{94 \cdot 4 \left(\dfrac{V_w}{V_s}\right)^{3/19} \left(1 + \dfrac{2c}{3\ell}\right) \ell^{11/19} V_s}{D_{eq}^{43/304}(Vol)^{0.47}d^{5/38}(Rc)^{27/19}} \quad \frac{mm^3}{min\text{-}N} \quad (6.30)$$

where

V_s = Wheel surface speed, m/min
c = Wheeldress depth on diameter, mm
ℓ = Wheel dress lead, mm/rev
Vol = Wheel bond percentage, percent
d = Grit size, mm
Rc = Workpiece Rockwell Hardness Number

The wheel bond percentage is given by

$$Vol = 1.33H + 2.2S - 8 \quad (6.31)$$

where

H = Wheel grade number and equals (0,1,2,3. etc) for hardness grades H,I,J,K . . . etc
S = Wheel structure number

and both may be obtained from the wheel designation. The equivalent diameter D_{eq} is a measure of the degree to which the wheel surface conforms to the workpiece surface. For external cylindrical grinding

$$D_{eq} = \frac{D_s D_w}{D_s + D_w} \quad (6.32)$$

where

D_s = Wheel diameter, mm
D_w = Workpiece diameter, mm

For internal grinding, the equivalent diameter is given by

$$D_{eq} = \frac{D_s D_w}{D_w - D_s} \quad (6.33)$$

For surface grinding, the equivalent diameter is equal to the wheel diameter D_s.

The following statements can be made, based on Eq. (6.29 – 6.33)

1. The cutting stiffness increases almost proportionately as the workpiece surface speed increases and decreases almost proportionately as the wheel surface speed increases.

2. An increase in the wheel diameter increases the equivalent diameter. The increase in equivalent diameter is greater for internal grinding as compared with external cylindrical grinding or surface grinding. The effect on the cutting stiffness is small, however, because of the small value of the exponent of D_{eq} in Eq. (6.30).

3. The cutting stiffness is lower with coarser dressing. Coarser dressing would result from larger values of the ratio c/ℓ or from larger values of the dress lead ℓ for a given value of c/ℓ.

4. The cutting stiffness increases proportionately with the width of grind, w.

Eq. (6.30) is valid only for grinding performed with freshly dressed wheels. As cumulative grinding time increases, flats develop on the abrasive grains, decreasing the interface stresses and hence the material removal parameter. If the wheel has a self-sharpening effect, the material removal parameter would also increase. It should also be noted that the Eq. (6.30) is applicable only to materials characterized as easy to grind.

The dependence of the cutting stiffness on the speed ratio V_w/V_s, however, seems to be a general result. Snoeys and Brown[8] cite experimental results by a number of researchers that confirm that, at the low values of V_w/V_s commonly used,

$$\frac{K_w}{w} \simeq k_c \frac{V_w}{V_s} \qquad (6.34)$$

where k_c is called the cuttability index and depends on the workpiece material, grinding wheel composition, dressing conditions, etc. The experimental results indicated that k_c varied from 100–1000 N/ m/mm for a wide range of combinations of workpiece materials and grinding wheels. Eq. (6.29) and (6.30) indicate that

$$\frac{K_w}{w} \sim \left(\frac{V_w}{V_s}\right)^{16/19} \qquad (6.35)$$

Inasaki and Yonetsu[11]

$$\frac{K_w}{w} \sim \left(\frac{V_w}{V_s}\right)^{\epsilon_0} (u_0)^{\epsilon_0^{-1}} \qquad (6.36)$$

where
   u_o = Mean depth of cut, μm
and ϵ_o = Grinding force exponent, $0 < \epsilon_o < 1$

Wheel wear stiffness—The wheel wear stiffness is related to the wheel removal parameter Λ_s by an equation similar to Eq. 6.29.

$$Ks = \frac{V_s w}{\Lambda_s} \qquad (6.37)$$

where
   K_s = Wheel wear stiffness, N/mm
   V_s = Wheel surface speed, m/min
and
   Λ_s = Wheel removal parameter, mm³/min-N

The wheel removal parameter is related to grinding conditions and details of the wheel dressing preceding the grinding operation. The wheel removal parameter increases with the normal grinding force intensity (force per unit width of cut) unlike the metal removal parameter[23]. That is a result of the fact that at low values of the grinding force intensity, attritious wear predominates; but at higher grinding force intensities grain fracture accounts for a greater proportion of wheel wear. Hahn and Lindsay[23] have formulated the following relationship, assuming single-point diamond dressing, for easy-to-grind workpiece materials. Data from 200 grinds were used to generate the empirical relationship.

$$\Lambda_s = \frac{k_a \ell^2 (1 + \frac{c}{\ell})\, F_n'\, V_s}{D_{eq}^{1.2/vol}(vol)^{0.85}} \qquad \frac{mm^3}{min\text{-}N} \qquad (6.38)$$

A similar empirical relationship has also been formulated[23] for difficult-to-grind materials.

$$\Lambda_s = \frac{k_b D_{eq}^{2/3}(F_n')^{5/3} V_s}{(Vol)^{1.86}} \qquad \frac{mm^3}{min\text{-}N} \qquad (6.39)$$

where
   F_n' = Normal grinding force per unit width, N/mm = Z_w'/Λ_w
   Z_w' = Stock removal rate per unit width, mm³/min-mm
and k_a, k_b = Constants dependent on the coolant and wheel grain type.
   The following statements can be made, based on Eq. (6.37)–(6.40):

(1) The wheelwear stiffness decreases as the normal grinding force intensity increases, such as at higher mean depths of cut.

(2) The wheelwear stiffness decreases almost proportionately as the workpiece surface speed increases and increases almost proportionately

as the wheel surface speed increases. That dependence is a result of the dependence of the wheelwear stiffness on the normal grinding force intensity.

(3) An increase in the wheel diameter increases the equivalent diameter. That effect is more marked in internal grinding as compared with surface or external cylindrical grinding. The wheelwear stiffness increases by a small amount for easy-to-grind materials, according to Eq. (6.38).

(4) The wheelwear stiffness increases proportionately with the width of cut, w.

(5) The wheelwear stiffness increases as the wheel bond percent, Vol, increases, such as for wheels with more open structure and fewer abrasive grains per unit volume of wheel material.

(6) The wheelwear stiffness decreases markedly as the workpiece material hardness increases. Higher workpiece material hardness results in a lower metal removal parameter value and a correspondingly higher value of the normal grinding force intensity for a given stock removal rate. That in turn results in a higher value of the wheel removal parameter and a lower wheelwear stiffness.

Relationships such as Eq. (6.37)–(6.40) are of great use because they indicate clearly the effect of grinding process parameters on the chatter loop parameter K_s, the wheelwear stiffness. The mechanisms of grinding wheelwear have also been studied at great length by other researchers under a great variety of grinding conditions [24-25]. Results of this and other related research should also be useful in determining the effect of grinding process parameters on the wheelwear stiffness.

Inasaki and Yonetsu[11] and Snoeys and Brown[8] relate the wheelwear stiffness to the cutting stiffness as indicated below

$$K_s = K_w \frac{G}{(V_w/V_s)} \tag{6.41}$$

where G = Grinding ratio

$$= \frac{\text{Volumetric workpiece material removal rate}}{\text{Volumetric wheelwear rate}}$$

Eq. (6.41) does not provide the same relationship between the grinding process parameters and the wheelwear stiffness that Eq. (6.37)–(6.40) provide. However, in view of the fact that grinding ratios vary typically from 5–100 and the speed ratios V_w/V_s vary from 1/100-1/20, it is clear that wheelwear stiffnesses are typically 2 to 4 orders of magnitude higher than cutting stiffnesses.[8]

Contact Stiffness — The contact stiffness K of the grinding wheel is very significant in determining chatter behavior. Under the following assumptions, an analytical expression for the contact stiffness can be determined.

(1) The grinding wheel is composed of discrete grains supported by springs with the same spring constant k.

(2) The grains and springs act independently of each other.

(3) The radial spacing between the grains is large with respect to the wheel surface deformation y_k and

(4) The probability density function describing the distribution of grains along the radial direction is directly proportional to radial distance from the wheel surface.

Snoeys and Brown[8] report that the contact stiffness K is then given by

$$K = 3.72w(k/a)^{0.25}(D_{eq})^{-0.25}(F_n')^{0.75} \tag{6.42}$$

$$= K_0 w D_{eq}^{0.25}(F_n')^{0.75}$$

where a = Grain spacing parameter in probability distribution function and K_0 = Wheel constant

Brown[6] has considered a number of other assumed probability density functions for grain distribution and given corresponding analytical expressions for the contact stiffness. The derived exponents of the equivalent diameter D_{eq} and the normal grinding force per unit width F_n' vary with the assumed probability density function. Inasaki and Yonetsu[11] have given an expression for the contact stiffness obtained using the Hertzian elastic contact theorem. The grinding wheel is assumed to be a homogeneous elastic disc and the abrasive grains are assumed to be spherical. In all of those cases, the contact stiffness can be seen to increase with the level of the normal grinding force per unit width.

Grinding wheel characteristics directly affect the parameters k and 'a' in Eq. (6.42) and hence the contact stiffness. There is another indirect effect, however, which is often more significant than the direct effect.[8] The grinding force depends on wheel characteristics, being higher for harder wheels than for softer wheels, other conditions remaining the same. Consequently, the wheel contact stiffness, which depends strongly on the grinding force level, is higher for harder wheels than softer wheels.

Brown[6] has determined experimentally the dependence of the contact stiffness on a number of parameters. The contact stiffness varied significantly when measured at different locations on the wheel, owing to local variations within the wheel. The average of those local contact stiffnesses around the wheel was therefore used to describe wheel contact stiffness. The experimental results indicated that the assumption concerning independence of grain deflections was appropriate. Also, measured contact stiffnesses on different wheels with the same grade, grain size, structure, and abrasive were close to each other, indicating that these wheel parameters have a dominant influence on contact stiffness. The measured contact stiffness did increase as the preload on the wheel increased. The exponent of the preload was 0.47. Snoeys and Brown[8] have reported a preload exponent obtained experimentally, of 0.37. Bartalucci and Lisini[9] have also determined experimentally that the contact stiffness increases with the load, but does so less than linearly.

Brown's results[6] indicated the following relationships also:

(1) The dressing condition had some effect on the contact stiffness. Diamond dressing resulted in higher K at low preload levels, as compared with crush dressing. At higher preload levels, the dressing conditions had little effect on contact stiffness.

(2) Grain size had some effect on the contact stiffness at low preloads. Finer grains resulted in lower contact stiffness. At higher preloads, grain size had little effect on contact stiffness.

(3) Wheel structure had a significant effect on contact stiffness, for low and high preloads. For a given preload, softer wheels resulted in higher contact stiffness than harder wheels.

It should be noted that the relationships for contact stiffness proposed above are not valid under all conditions. Baylis and Stone[27] have reported that the contact stiffness is not proportional to width but tends to an asymptotic value, for Borazon grinding wheels specially designed to have wheel flexibility. The subject of elastic wheel deflection in grinding continues to be an area of active research. Consequently effects ignored by earlier analyses, such as the effect of tangential force components on elastic deflections of grains, are being studied more carefully[28] resulting in better understanding of the mechanisms involved. When this research produces quantitative reationships similar to Eq. (6.42), new or modified conclusions concerning grinding chatter can be formulated.

Miyashita et al[5] have reported that contact stiffnesses of the grinding wheel as well as the control wheel are significant factors governing chatter in centerless grinding. Control wheel contact stiffnesses were seen to be lower than grinding wheel contact stiffnesses.

Contact Length — The finite contact length ℓ_c serves to average out and attenuate the effect of higher frequency surface undulations on the wheel and workpiece. Since the workpiece surface speed is usually much lower than the wheelspeed, workpiece surface undulations are attenuated more than wheel surface undulations, reducing the feedback of workpiece surface undulations and resulting in the phenomenon of wheel regenerative chatter at low workpiece speeds.

The real contact length in cylindrical plunge grinding operations has been investigated by Verkerk[29] and others[30]. The real contact length in cylindrical plunge grinding was investigated for a variety of grinding conditions including different workpiece speeds, wheel particle grain sizes, wheel hardnesses and specific metal removal rates[29]. The real contact length was always greater than the geometrical contact length ℓ_g, given by

$$\ell_g = \sqrt{y_c D_{eq}} \qquad (6.43)$$

where y_c = Depth of cut

The difference was ascribed to the elastic deformation of the contact zone in the presence of the grinding force. The real contact length was found to be anywhere from 1 to 3 times the geometrical contact length depending on the grinding conditions[29]. For a fixed depth of cut, the real contact length was greater at higher ratios of workspeed to wheelspeed. That is a result of the higher material removal rates and larger grinding forces. The wheel abrasive grain size had little effect on the contact length. Wheel hardness, as indicated by wheel grade, had some effect, especially for workpiece materials with lower hardness. Softer wheels resulted in greater elastic deformation of the wheel and hence higher contact lengths. Other experimental work summarized by Saini[30] confirms the observations noted above. Contact lengths in centerless grinding owing to grinding wheel and control wheel flexibility were measured by Miyashita et al.[40] and seen to be significant factors affecting chatter behavior.

The relationships noted above are qualitative. Quantitative relationships between the real contact length and grinding conditions are not available.

Machine Dynamic Compliance — The dynamic compliance of the grinding machine is largely determined by the details of the machine structure design. It is the only chatter loop parameter over which the machine tool manufacturer has any control. It should be clear from Eq. (6.8) and (6.10) that higher machine static stiffness K_m will reduce susceptibility to chatter. Similarly, reduction of the magnitude of the negative real part of the machine dynamic compliance $G_m(j\omega)$ will reduce machine susceptibility to chatter in general. The effect of machine dynamic compliance on chatter behavior is more complex than the relationships stated above and is represented more properly by the regeneration spectrum concept. However, the relationships stated above are simpler and are used more by machine tool manufacturers.

The simplest form of the machine dynamic compliance is given below and is appropriate if 1 mode of vibration is dominant over the others.

$$\frac{G_m(s)}{K_m} = \frac{1/K_m}{\dfrac{s^2}{\omega_m{}^2} + \dfrac{2\zeta_m}{\omega_m}s + 1} \tag{6.44}$$

where ω_m = Natural frequency of single mode of vibration
ζ_m = Damping ratio for the single mode of vibration.

For a single mode of vibration, the real part of $G_m(j\omega)$ /K_m is positive for all frequencies below the natural frequency ω_m and negative for all frequencies higher than ω_m. Also, the lower the damping ratio ζ_m, the more negative the real part.

It is more general, however, to find more than 1 mode of vibration to be significant. The different modes of vibration would involve different components of the machine to different extents in the vibration. For example, lower frequency vibrational modes would involve large masses such as support components for the roll workpiece in roll grinding machines. Alternatively, bending vibration of long slender rolls could give rise to low frequency modes. Higher frequency modes involve smaller masses and could correspond to rocking of the grinding wheel on the spindle or bending vibration of the wheel spindle. For such cases, the machine dynamic compliance can be represented as

$$\frac{G_m(s)}{K_m} = \sum_{i=1}^{n} \frac{1/K_i}{\dfrac{s^2}{\omega_i{}^2} + \dfrac{2\zeta_i}{\omega_i}s + 1} \tag{6.45}$$

where K_i = Modal stiffness of ith mode
ω_i = Natural frequency of ith mode
ζ_i = Damping ratio of ith mode
and   n = Number of modes

Snoeys and Brown[8] have tabulated static stiffnesses of representative universal grinding machines. Plane grinders ranging from 1 to 15 HP were considered and had static stiffnesses ranging from 18–80 N/μm. Brown investigated machine static stiffnesses of approximately 20 grinding machines, primarily steel mill and aluminum foil roll grinders. The static stiffnesses ranged from 50–90 N/μm. The dyanmic compliance of grinders of the same type but manufactured by different companies were also seen to be remarkably similar. Machine static stiffness in internal grinding is determined primarily by the quill or wheel spindle. It is determined by the geometry of the part being ground and is usually lower than static stiffnesses for plane or cylindrical grinders. Brown[6] has reported a static stiffness of 2-5 N/μm for an internal grinder investigated, and Hahn[31] has reported static stiffnesses varying from 4–9 N/μm.

The negative real part of the normalized compliance $G_m(j\omega)$ depends on the damping in the machine structure. The lower is the damping, the higher in magnitude or the more negative is the real part of $G_m(j\omega)$. Snoeys and Brown[8] suggest the following range as representative.

$$-10 \leq \text{Min}_\omega \quad (\text{Re}(G_m(j\omega))) \leq -1 \qquad (6.46)$$

The left-hand side limit corresponds to very lightly damped machine structure and the right-hand side to a well-damped machine structure. The factors governing machine structure dynamic compliance are obviously many and are described more fully in references on machine tool design[32] and vibration analysis. The implication of any machine structure design or modification for chatter behavior, is, however, evaluated properly only in conjunction with the chatter stability analysis techniques described in the body of this chapter.

Nomenclature

a	Grain spacing parameter
c	Wheel dress depth on diameter
D_{eq}	Equivalent diameter
D_s	Wheel diameter
D_w	Workpiece diameter
d	Grit size on wheel
F_c	Grinding force
F_n'	Normal grinding force intensity
G	Grinding ratio
G(s)	Open loop transfer function
$G_m(s)$	Normalized machining system dynamic compliance
H	Wheel grade number
h(s)	Geometrical rounding effect function
Im	Imaginary part of complex number
j	$\sqrt{-1}$
K	Contact zone stiffness
K_{cc}	Contact zone stiffness of control wheel
K_{cs}	Contact zone stiffness of grinding wheel
K_i	Modal stiffness of ith mode of vibration

K_m Static stiffness of machine
K_o Constant in contact zone stiffness equation
K_s Wheelwear stiffness
K_w Cutting stiffness
k Grain spring constant
k_a, k_b Constants in wheel removal parameter equations
k_c Cutability index
ℓ Wheel dress lead
ℓ_c Real contact zone length
ℓ_g Geometrical contact zone length
N_c Control wheel rotational speed
N_s, N_{si} Grinding wheel rotation speeds
N_w Workpiece rotational speed
n_s Number of lobes on wheel periphery
$P(s), Q(s)$ Polynomials in s
$R(\omega)$ Regeneration spectrum
Rc Rockwell hardness number
Rc Real part of complex number
$R_s(s)$ Wheel regeneration spectrum
$R_w(\omega)$ Workpiece regeneration spectrum
S Wheel structure number
s Laplace variable
T Time period of revolution
u Total infeed
u_o Depth of cut per revolution
V_s Wheel surface speed
V_w Workpiece surface speed
w Width-of-cut
y_k Contact area deformation
y_m Machining system deflection
$Z_s(s)$ Wheel contact length filter transfer function
$Z_w(s)$ Workpiece contact length filter transfer function
$Z_{wc}(s)$ Control wheel contact length filter transfer function
$Z_{ws}(s)$ Grinding wheel contact length filter transfer function
Z_w' Stock removal rate per unit width
α Real part of $s = \alpha + j\omega$
α_1 Angular orientation of control wheel
α_{max} Maximum real part of chatter loop poles
β Angular orientation of grinding wheel
γ Center-height angle
δ_s Total wheelwear
δ_w Total workpiece depth-of-cut

ϵ_o Exponent in cutting force relation
ϵ, ϵ' Parameters in geometrical rounding effect function
ζ_i Damping ratio of ith mode
ζ_m Damping ratio of machine vibrational mode
θ Workplate angle
Λ_s Wheel removal parameter
Λ_w Metal removal parameter
μ_s Overlap factor for wheel
μ_w Overlap factor for workpiece
ν_{si} Fraction of lobe on wheel periphery at chatter frequency$_i$
τ_s, τ_{si} Wheel periods of revolution
τ_w Workpiece period of revolution
τ_1, τ_2 Time delays in centerless grinding
ϕ_1, ϕ_2 Orientation of workplate and control wheel contact zones relative to grinding wheel contact zone
ω Frequency
ω_i Natural frequency of ith mode of vibration
ω_m Natural frequency of machine vibration
ω_{max} Frequency corresponding to peak of regeneration spectrum

References

1. Kegg, R.L., "Industrial Problems in Grinding," *Annals of the CIRP*, Vol. 32/2, pp. 559-61, 1983.

2. Hahn, R.S., "The Influence of Grinding Machinability Parameters on the Selection and Performance of Precision Grinding Cycles," Proceedings of the International Conference on Machinability Testing and Utilization of Machining Data, American Society of Metals, Metals Park, Ohio, pp. 164–78, Sept. 1978.

3. Maris, M., Snoeys, R., and peters, J., "Analysis of Plunge Grinding Operations," *Annals of the CIRP*, Vol. 24/1, pp. 225–30, 1975.

4. Weck, M., and Schiefer, K.H., "Interaction of the Dynamic Behavior Between Machine Tool and Cutting Process for Grinding," *Annals of the CIRP*, Vol. 28/1, pp. 281–85, 1979.

5. Miyashita, M., Hashimoto, F., and Kanai, A., "Diagram for Selecting Chatter Free Conditions of Centerless Grinding," *Annals of the CIRP*, Vol. 31/1, pp. 221–23, 1982.

6. Brown, D.L., *Grinding Dynamics*, Ph.D. Thesis, Department of Mechanical and Industrial Engineering, The University of Cincinnati, 1976.

7. Hahn, R.S., "Grinding Chatter in Precision Grinding Operations—Causes and Cures," Technical Paper No. MR78-331, Society of Manufacturing Engineers, 1978.

8. Snoeys, R., and Brown, D., "Dominating Parameters in Grinding Wheel and Workpiece Regenerative Chatter," Proceedings of the 10th International

Machine Tool Design and Research Conference, pp. 325–48, Pergamon Press, 1969.

9.   Bartalucci, B., and Lisini, G.G., "Grinding Process Instability," Transactions of the ASME, *Journal of Engineering for Industry*, Vol. 91, pp. 597–606, August 1969.

10.   Verkerk, J., "The Real Contactlength in Cylindrical Plunge Grinding," *Annals of the CIRP*, Vol. 24/1, pp. 259–64, 1975.

11.   Inasaki, I., and Yonetsu, S., "Regenerative Chatter in Grinding," Proceedings of the 18th International Machine Tool Design and Research Conference, pp. 423–29, Pergamon Press, 1977.

12.   Thompson, R.A., "The Character of Regenerative Chatter in Cylindrical Grinding," Transactions of the ASME, *Journal of Engineering for Industry*, Vol. 95, pp. 858–64, August 1973

13.   Thompson, R.A., "On the Doubly Regenerative Stability of a Grinder," Transactions of the ASME, *Journal of Engineering for Industry*, Vol. 96, pp. 275–80, February 1974.

14.   Thompson, R.A., "On the Doubly Regenerative Stability of a Grinder: The Combined Effect of Wheel and Workpiece Speed," Transactions of the ASME, *Journal of Engineering for Industry*, Vol. 99, pp. 237–41, February 1977.

15.   Dorf, R.C., *Modern Control Systems*, Addison-Wesley Publishing Co., Reading, Mass., 1967.

16.   Scientific Subroutine Package, International Business Machines Corporation, New York.

17.   User's Manual, International Mathematical and Statistical Library (IMSL), Houston, Tex.

18.   Srinivasan, K., "Application of the Regeneration Spectrum Method to Wheel Regenerative Chatter in Grinding," Transactions of ASME, *Journal of Engineering for Industry*, Vol. 104, pp. 46–54, February 1982.

19.   Srinivasan, K., "Application of the Regeneration Spectrum Method to Workpiece Regenerative Chatter in Grinding," Proceedings of the 9th North American Manufacturing Research Conference, pp. 283–89, University Park, Penn., May 1981.

20.   Merritt, H.E., "Theory of Self-Excited Machine-Tool Chatter. Contribution to Machine-Tool Chatter Research—1," Transactions of the ASME, *Journal of Engineering for Industry*, Vol. 87, pp. 447–54, November 1965.

21.   Hahn, R.S., and Lindsay, R.P., "Principles of Grinding—Part I: Basic Relationships in Precision Grinding," Machinery, pp. 55–62, July 1971.

22.   Hahn, R.S., and Lindsay, R.P., "Principles of Grinding—Part II: The Metal Removal Parameter," Machinery, pp. 33–39, August 1971.

23.   Hahn, R.S., and Lindsay, R.P., "Principles of Grinding—Part III: The Wheel Removal Parameter," pp. 33–38, September 1971.

24.   Malkin, S., and Cook, N.H., "The Wear of Grinding Wheels: Part I. Attritious Wear," ASME Transactions, *Journal of Engineering for Industry*, Series B, Vol. 93, pp. 1120–1128, 1971.

25.   Malkin, S., and Cook, N.H., "The Wear of Grinding Wheels: Part 2. Fracture Wear," ASME Transactions, *Journal of Engineering for Industry*, Series B, Vol. 93, pp. 1129–1136, 1971.

26.   Rubenstein, C., "The Mechanics of Grinding," *International Journal of Machine Tool Design and Research*, Vol. 92, pp. 127–34, 1972.

27.   Baylis, R.J., and Stone, B.J., "The Build Up and Decay of Vibration During Grinding," *Annals of the CIRP*, Vol. 32/1, pp. 265–68, 1983.

28.   Saini, D.P., Wager, J.G., and Brown, R.H., "Practical Significance of Contact Deflections in Grinding," *Annals of the CIRP*, Vol. 31/1, pp. 215–19, 1982.

29.   Verkerk, J., "The Real Contact Length in Cylindrical Plunge Grinding," *Annals of the CIRP*, Vol. 24/1, pp. 259–64, 1975.

30.   Saini, D.P., "Elastic Deflections in Grinding," *Annals of the CIRP*, Vol. 29/1, pp. 188–94, 1980.

31.   Hahn, R.S., "The Fundamentals of Precision Grinding," SME Paper No. MR76–370, Society of Manufacturing Engineers, 1976.

32.   Koenigsberger, F., and Tlusty, J.F., *Machine Tool Structures*, Pergamon Press, 1970.

33.   Hahn, R.S., "Vibration Problems and Solutions in Grinding," ASTME Paper No. MR69–246, 1969.

34.   Tonshoff, H.K., "Process Control in Internal Grinding," *Annals of the CIRP*, Vol. 29/1, pp. 207–11, 1980.

35.   Thompson, R.A., "The Dynamic Behavior of Surface Grinding. Part I—A Mathematical Treatment of Surface Grinding," Transactions of the ASME, *Journal of Engineering for Industry*, pp. 485–91, May 1971.

36.   Thompson, R.A., "The Dynamic Behavior of Surface Grinding. Part 2—Some Surface Grinding Tests," Transactions of the ASME, *Journal of Engineering for Industry*, pp. 492–497, May 1971.

37.   Malkin, S., "Review of Materials Processing Literature—1969–1970. Part 2—Grinding," Transactions of the ASME, *Journal of Engineering for Industry*, pp. 299–06, February 1973.

38.   Miyashita, M., "Chatter Vibration in Centerless Grinding," Bulletin of the Japanese Society of Precision Engineering, Vol. 3, No. 3, pp. 53–58, March 1969.

39.   Furukawa, Y., Miyashita, M., and Shiozaki, S., "Chatter Vibration in Centerless Grinding. Research I: Work Rounding Mechanisms Under the Generation of Self-Excited Vibration, *Bulletin of the Japanese Society of Mechanical Engineers*, Vol. 13, No. 64, pp. 1274–1283, 1970.

40.   Furukawa, Y., Miyashita, M., and Shiozaki, S., "Vibration Analysis and Workrounding Mechanism," *International Journal of Machine Tool Design and Research*, Vol. 11, Pergamon Press, 1971.

41.   Harris, C.M., and Crede, C.E., *Shock and Vibration Handbook*, Second Edition, McGraw-Hill, New York, 1976.

42.   Cegrell, G., "Variable Wheel Speed—A Way to Increase the Metal Removal Rate," Proceedings of the 13th International Machine Tool Design and Research Conference, pp. 653–58, University of Manchester, Pergamon Press, 1972.

43.   Pahlitzsch, G., and Cuntze, E.O., "Reduction of Chatter Vibration During Cylindrical and Plunge Grinding Operation," Proceedings of the 6th International Machine Tool Design and Research Conference, University of Manchester, Pergamon Press, 1965.

CHAPTER **7**

Precision Grinding Cycles

Robert S. Hahn

Introduction

Grinding cycle is the term used to designate the input parameters imposed on a grinding machine to produce the desired output. Those machine input parameters—viz., wheel speed, work speed, feedrate, etc.—were discussed in Chapter 1 and are illustrated in Fig. 1.2. The output variables on the finished workpiece—viz., size, taper and form errors, surface finish deviations, thermal damage and cycle time—are also illustrated in Fig. 1.2. The selection of the proper combination of machine input parameters to produce the desired output often requires considerable expertise, especially when high-quality output is desired in a short cycle time. The grinding of the bore of automotive valve lifters at high production rates to stringent requirements on size, taper and surface finish illustrates the difficulty in obtaining fast production cycles while maintaining close tolerances. A reduction in cycle time of 1 second is worth about $1,000,000 per year. However, the quality deteriorates if the cycle is sped up beyond a certain limit. This chapter deals with the various factors governing the grind cycle input parameters and their relation to output quality.

Rigidity

The rigidity of grinding machines plays an important role in controlling size, taper, and form errors. The normal force induced between the

170

grinding wheel and workpiece causes the system to deflect. The grinding system consists of a large number of elastic elements in series. As an operator turns a hand wheel or a CNC, command is fed to a servo-drive motor to cause an infeed motion, all the elements in the force loop deflect. The feed ball screw, its thrust bearings, the ball nut, the feed-slide structure, the wheelhead housing, wheelhead spindle and bearings, the workpiece and its supporting structure contribute to the total deflection. The stiffness or rigidity of each of those elements is measured by its spring constant K (N/m). The overall stiffness K_t of the entire system is found by adding reciprocally the stiffness of each element:

$$\frac{1}{K_t} = \frac{1}{K_1} + \frac{1}{K_2} + \frac{1}{K_3} + \text{- - - -} + \frac{1}{K_a} \qquad (7.1)$$

where K_1, K_2 --- are the spring constants of each element, and K_a is the incremental spring constant of the abrasive wheelwork contact region. Note that the stiffness of the feed ball screw, ball nut and thrust bearings does not affect slide positioning when linear feedback scales are used for slide positioning. Also note that the element with the lowest rigidity (usually the workpiece on OD grinders or the quill and spindle on ID grinders) dominates the overall system rigidity and that increasing the stiffness of other elements in excess of 10 times the least rigid element contributes little to the system stiffness.

The stiffness of the grinding wheel K_a, owing to deflection in the wheelwork contact zone, is governed by the modulus of elasticity of the abrasive wheel and is highly nonlinear. Since the modulus for vitrified wheels is 1/5 or less than that for steel, the wheel is relatively more compliant than steel. A typical relation between contact stiffness and normal force intensity for a 2A60K6VLE wheel with D_e = 4.93 in. is[1]:

$$K_a' = K_o \ D_e^{.25} \ (F_n')^{.75} \qquad (7.2)$$

where K_a' in the contact stiffness per inch of width:

K_o = .58 x 10^5
D_e = equivalent diameter (in.)
F_n' = normal interface force per inch of width (lb/in)

In addition to the deflection normal to the wheelwork interface a tangential component exists that is generally of no consequence in grinding. However, angular deflections also take place and can be very important where the grinding wheel is used to grind a shoulder or deep slot, as illustrated in Fig. 1.12d. The angular stiffness κ and the lateral stiffness K of a cantilever quill are illustrated in Fig. 1.11a.

Automatic Sizing

Precision grinding requires that a succession of workpieces be ground to a given size tolerance. It is often quite difficult to produce a large number of parts all within a narrow tolerance. Two methods are generally employed to control size. In-process gaging is used in many cases where space is available to apply electronic gage fingers that ride on the surface being ground and generate a signal to terminate the grinding cycle when the part reaches the desired size. A second method uses the dresser diamond or dressing device as a reference position for sizing. That "diamond sizing" method is applicable where it is inconvenient or impossible to apply in-process gaging, such as in blind holes. In the simplest case of diamond sizing the dresser diamond is mounted on the work-supporting element, axially displaced from the grinding area and located at a specific radial distance from the work centerline equal to the desired size. In this case, once the wheel has been dressed at a specific position of the feed slide, it is only necessary to bring the feed slide back to that position to grind the workpiece to the proper size. That is illustrated in Fig. 7.1.

Note that gradual thermal expansion of the structure supporting the dresser diamond will cause a corresponding change in size. Accordingly, that "diamond-work centerline" loop should be kept as small as possible or temperature controlled. Thermal expansions in other parts of the machine have no effect on size as long as the wheel is dressed by the diamond and immediately grinds the workpiece. Size drift resulting

Fig. 7.1   Diamond-work centerline thermal expansion loop.

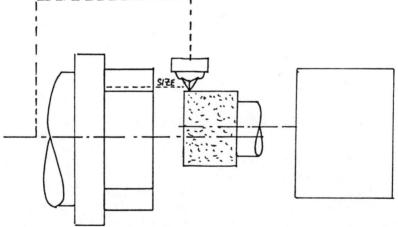

from thermal expansion may occur on subsequent grind cycles until a dressing operation restores size. On subsequent grind cycles without dressing, wheelwear will also cause a gradual drift in size which, again, will be restored by the dressing operation. Wear of the dresser diamond is a third cause of systematic size drift.

Sometimes the dressing device is mounted behind the wheel and not on the work-supporting element. In such cases an additional dresser slide is required to feed the dressing device into the wheel to compensate for wheelwear; whereas, in the former case, the feed slide was used to feed the wheel for grinding and, also, for "compensation" for wheelwear. Note that when a dressing operation is invoked, the position of the slide must be incremented by an amount equal to the amount of wheelwear plus the diamond depth of penetration into the wheel. That is called the "compensation." The position of the feed slide, when one is grinding to size, must also be incremented by the same amount. Note also that on workpieces having large stock variations (stock allowance), the amount of wheelwear may also vary, and for a fixed compensation, that results in a variation in the diamond depth of penetration. That, in turn, produces a variation in the sharpness of the wheelface and surface finish. On production grinding cycles, thermal expansion, diamond wear, and wheelwear usually cause a unilateral systematic drift in size in contrast to random bilateral piece-to-piece size variations about a mean.

These apparently random errors may be caused by malfunction of the machine-slide positioning system or dresser system. However, on good-quality grinding machines, the slide movements often repeat within a few microns (10^{-6}m or 40 micro inches) but this excellent machine repeatability does not guarantee the same degree of precision in the ground workpiece. The cause of many so-called random errors correlates with variations in the elastic deflection of the machine-wheel workpiece system. They are discussed below.

Grinding Force Profiles

Production grinding machines generally consist of a rotating, work-holding device, into which workpieces are automatically loaded and unloaded, and a grinding wheelhead mounted on a cross slide, which is capable of moving the wheel radially into contact with the workpiece. Fig.7.2 shows a record of the radial cross-slide displacement (upper trace) during 2 consecutive grinding cycles of a production grinding machine[2]. The lower trace is a record of the normal interface force existing at the wheelwork contact area during the grind cycle as obtained by a

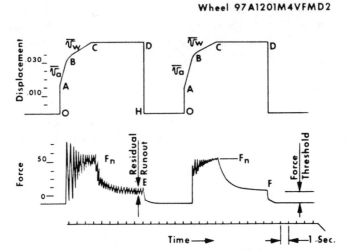

Fig 7 2 Typical grinding cycle showing cross-slide position, and normal grinding force vs. time. Workpiece: automotive, double-groove waterpump bearing. Operation: grind ball tracks.

force transducer. In a controlled-force grinding machine, illustrated in Fig. 7.3, the cross slide is moved at a high velocity from O to A (in Fig. 7.2) by the applied force F_a shown in Fig. 7.3, bringing the wheel close to the unground running-out workpiece. At point A in Fig. 7.2, the dashpot C, shown in Fig. 7.3, is engaged, causing the slide to approach the workpiece at the approach velocity $\bar{v}_a$, which is, simply:

$$\bar{v}_a = \frac{F_a}{C} \tag{7.3}$$

As the cross slide moves from A to B (Fig. 7.2), the wheel picks up the running-out workpiece. Large instantaneous force pulsations may occur on the wheel during the early stages of rounding up, as shown by the first force trace. As rounding up takes place, the force pulsations reduce and tend toward an average grinding force F_n according to:

$$F_n = F_a - C\bar{v}_w \tag{7.4}$$

where $\bar{v}_w$ is the plunge-grinding velocity. The majority of the stock is removed and further roundup accomplished from B to C.

At C, the cross slide strikes a stop (Fig. 7.3) and comes to rest. However, the grinding wheel continues to grind because of residual spindle deflection and sparks out during the interval CD. At D, the slide is re-

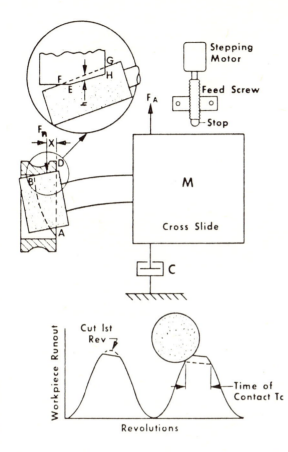

Fig. 7.3 Wheelwork engagement and rounding up for internal grinding of ball-bearing inner ring on controlled-force grinder.

tracted and the workpiece, one hopes, is at size with the proper surface finish, roundness, and microprofile. The work is unloaded and a new workpiece reloaded from H to O in preparation for the next cycle.

Feedrate grinding machines, in contrast with controlled-force machines, feed the cross slide at a prescribed feedrate $\bar{v}_f$, gradually developing grinding force as the deflection of the system increases. Fig. 7.4 shows the cross-slide position (line OAB) and the position of the cutting surface of the wheel (line OEF) during a simple feed and sparkout cycle. The vertical distance between line OAB and line OEF is the deflection in the system that is proportional to the grinding force.

During the sparkout period AB, the force (and deflection) decays to a small but finite "threshold force" or deflection F_{th}/K_m.

Stock variations of incoming parts often cause size variations in the finished parts if the steady-state force has not been achieved in the rough-grind portion of the cycle, as illustrated in Fig. 7.5, where the first

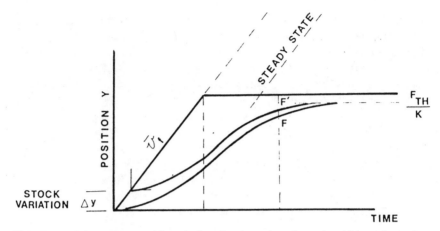

Fig. 7.4   Cross-slide position during feed motion O to A with sparkout A to B and cutting surface of wheel O to E to F for workpieces having large and small stock allowance.

part had more stock than the second part. The value of the force in the system at the termination of the cycle governs the final size and surface finish (unless in-process gaging is used). Consequently, the size envelope produced may depend upon initial stock variations and variations in threshold force which, in turn, depend upon the wheel sharpness variations and changes in the conformity D_e.

The Rounding-Up Process

In a typical grinding cycle, the first contact with the workpiece is usually made at the high spot on the work. During the first few revolutions of the workpiece, the wheelwork contact is intermittent. As the high spot

Fig. 7.5   Normal force profile in a typical feedrate cycle for workpieces with large and small stock allowance.

FEED RATE GRIND

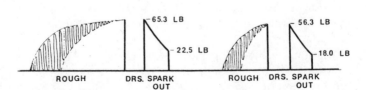

engages the grinding wheel, the wheel spindle deflects both laterally and angularly. That tends to concentrate the grinding force on the inner edge of the grinding wheel.

In early stages of rounding up, the instantaneous width-of-cut, w, varies, as shown in Fig. 7.3 as the zone ABD. The instantaneous grinding force at the high spot, F_n, lies at the position x from the right-hand edge of the work. The instantaneous wheel depth-of-cut, h(x), is also shown in Fig. 7.3. In order to predict grinding behavior during the rounding-up process, equations describing the dynamic behavior of the grinding machine have been combined with grinding-process equations in a computer program called TRUFI[3]. That program prints out for each work revolution, the progressively diminishing workpiece runout; the peak-force intensity existing on the grinding wheel at the high point, the peak force itself, and the amount of wheelwear. Also given are the amount of stock required to true up the workpiece to a given accuracy and the time required to round up the workpiece.

The grinding-process Eq. (1.13) and (1.16) can be adapted to give instantaneous values of wheel depth-of-cut and wheelwear over contact regions with nonuniform force intensity distributions.

In order to verify the computer program, a production grinder was equipped with instrumentation to record the instantaneous peak grinding force and the instantaneous workpiece runout during the grinding of No. 3920 taper roller-bearing cups (4.4375 in OD or 112.5mm)[3].

Fig. 7.6 shows peak grinding forces on a controlled-force machine that reach a maximum of 300 lbs. (1130N) after 1 second followed by a decline to around 100 lbs. (444.N) after 6 seconds. The instantaneous runout is shown in the lower part of the figure, where an initial runout of .008 in. (.2mm) is reduced to less than .001 in. (.025mm) in about 6 seconds. Measured values of runout and force as well as computed values are plotted on semilog paper for comparison in Fig. 7.7. It will be seen that the agreement between computed and measured values is reasonable.

Workpieces flowing to a production grinder usually have a joint random distribution of stock allowance, runout, and taper. A sampling of workpieces is shown plotted in Fig. 7.8. Workpieces with a large amount of stock and small runout are easy to round up and convert into a good finished product. Workpieces with small amounts of stock and relatively large runout are difficult to round up and make into a good finished product. By using the rounding-up computer program, one can find the amount of stock required to round up a given initial runout when one is grinding under a given force or feedrate. Doing that for several values of initial runout, one can draw the force lines or corresponding feedrates shown in Fig. 7.8, which indicate the amount of stock required to correct a given runout.

Fig. 7.6 The rounding up during controlled-force grinding of the roller track of No. 3920 tapered roller-bearing bearing cups. The lower trace gives the instantaneous runout; shows the period of intermittent contact and the beginning of continuous contact at 5.6S. The middle trace shows the cross-slide advance. The top trace gives the instantaneous force pulses occurring on the grinding wheel, and having a steady-state force of 100 lbf.

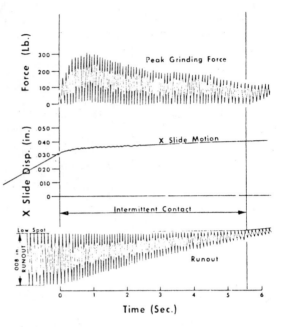

Fig. 7.7 Instantaneous runout and normal force vs. no. of work revolutions for controlled-force internal grinding of No. 3920 roller-bearing cups (4.4375 in. O.D.). Comparison of computed values with experimentally measured values.

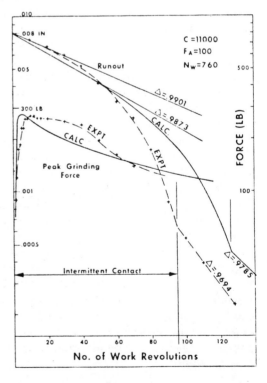

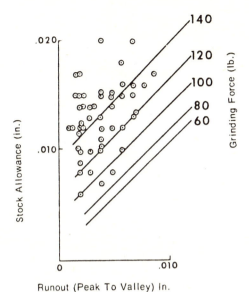

Fig. 7.8   Joint distribution of stock allowance and initial runout. The inclined lines give the stock required to true-up the runout at various grinding forces or corresponding feedrates. Workpiece: Universal joint cup

If a workpiece falls below a given force line, the stock allowance will be used up before the runout is corrected. For the case shown in Fig. 7.8, all workpieces will be rounded up if ground under 90 lbs (400N) of force or less, or the corresponding feedrate from Eq. (1.13). For 100 lbs (444N) of grinding force, there will be 2 parts (the 2 points below the 100-lb. line) that will not round up. Charts such as that shown in Fig. 7.8 are useful for selecting the highest permissible grinding force or feedrate.

The computer program TRUFI also outputs the amount of wheelwear occurring during the roundup process. That is an important variable, since it determines the amount of infeed, or so-called compensation, of the diamond required to true the grinding wheel. If the "compensation" is set too large, the wheel will be rapidly consumed and grinding cost per part will rise because of increased abrasive cost and machine downtime.

If the compensation is set too low, failure to dress will occur occasionally and surface finish, size, and microprofile will deteriorate. If the grinding force or feedrate is set too low in order to protect and conserve the wheel, cycle times tend to be long and, again, grinding costs go up. Consequently, it is important to select the grinding force/feedrate so that cycle times are short as possible without causing excessive wheel breakdown. For further discussion of rounding up, see Chapter 15 on adaptive control.

Influence of Threshold Force on Roundness

Workpieces often arrive at the grinding machine with some degree of out-of-roundness or eccentricity, which must be corrected by the grinding operation. The workpiece shown in Fig. 7.9, for example, has an eccentric stock distribution with the stock AB at the High Spot A and stock CD at the Low Spot D, giving a "total indicator reading" TIR of OD-OA. It is desired to grind the bore to diameter BC. If that part has no threshold force, a round hole will result after feeding the grinding wheel to "size" and sparking out. However, if there is a threshold force and perfect roundness has not been achieved in the rough grind, the spark-out process may leave some out-of-roundness. The maximum residual out-of-roundness, RO_{max}, is given by:

$$RO_{max} = F_{th}/K_t \qquad (7.5)$$

Fig. 7.9   Rounding up of eccentric stock under influence of threshold forces. Desired final diameter—CB; stock at high spot—AB; stock at low spot—CD

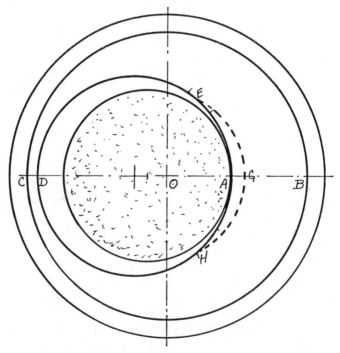

The rounding-up process where threshold forces exist can be analyzed as shown below.

The wheel depth of cut h, at the "high spot" A, will be zero until the feedrate $\bar{v}_f$ has generated sufficient force to exceed F_{th} and thereafter:

$$h = \frac{\Lambda_w(F_n - F_{th})}{\pi D_w N_w W} \qquad (7.6)$$

The "cutting stiffness" K_c can be obtained by differentiating Eq. 7.6 to obtain:

$$K_c = \frac{dF_n}{dh} = \frac{V_w W}{\Lambda_w} \qquad (7.7)$$

After some feed has taken place, the wheel will strike the work at point E and would sweep out the "unstressed locus" EGH if no deflection took place. However, some deflection does occur, so that the wheel sweeps out a deflected locus lying between the current work surface and the "unstressed locus." During the period of intermittent contact, grinding on the "high spot" takes place in a manner similar to a surface grinder. The force F_n, between wheel and work at the "high spot," progressively increases, reaching a steady state after 3 "Time Constants" τ_0. It has been shown[5] that the time constant of a plunge-grinding operation is:

$$\tau_0 = \frac{1}{\dfrac{K_t}{W} \left(\dfrac{\Lambda_w}{\pi D_w} + \dfrac{\Lambda_s}{\pi D_s} \right)} \qquad (7.8)$$

When the wheelwear rate is negligible ($\Lambda_s < \Lambda_w$), that reduces to:

$$\tau_0 = \frac{K_c}{K_t N_w} \qquad (7.9)$$

After 3 τ_0 seconds, both F_n and h reach a steady state on the High Spot.

After feeding a distance equal to TIR, the wheel will begin to strike the low spot, and the period of continuous contact begins. Once again, feed has to continue to build up the force on the Low Spot until it exceeds F_{th}. At that point the hole is considerably out-of-round. *The criterion for roundness is that the steady state be achieved on the low spot also.* If the spark-out commences before the steady state on the Low Spot has been attained, a residual out-of-roundness will result, since no more corrective action occurs after the force drops below F_{th}.

The stock that must be removed from the low spot to ensure roundness S_r is:

$$S_r = \int_0^{3\tau_0} \bar{v}_w dt \qquad (7.10)$$

where instantaneous plunge velocity of the wheel into the work $\bar{v}_w$ is given by Eq. (1.13).

$$\bar{v}_w = \frac{\Lambda_w}{\pi D_w} \frac{\left[1-e^{\frac{-K_t}{W}\left(\frac{\Lambda_w}{\pi D_w} + \frac{\Lambda_s}{\pi D_s}\right)t}\right]}{\left[\frac{\Lambda_w}{\pi D_w} \cdot + \frac{\Lambda_s}{\pi D_w}\right]} \bar{v}_f \qquad (7.11)$$

Using Eq. (7.11) and integrating Eq. (7.10) gives:

$$S_r = \frac{2.049\tau_0\bar{v}_f}{\left[1 + \frac{\Lambda_s D_w}{\Lambda_w D_s}\right]} \qquad (7.12)$$

When the wheelwear rate is negligible ($\Lambda_s < \Lambda_w$), that reduces to:

$$S_r = 2.049 \; \tau_0 \; \bar{v}_f \qquad (7.13)$$

If the stock available on the low spot is S_a, the optimum feedrate $\bar{v}_f$, to round up the part in the least time, is:

$$\bar{v}_{fopt} = \frac{S_a}{2.049\tau_0} = \frac{S_a K_t N_w}{2.049 K_c} \qquad (7.14)$$

The time required to round up T_{rup} is:

$$T_{rup} = T_{int} + T_{fth} + 3\tau_0 \qquad (7.15)$$

where T_{int} = time of intermittent contact
      = $TIR/\bar{v}_f$
      T_{fth} = time to develop F_{th} on Low Spot
      = $F_{th}/K_t v_f$

Eq. (7.14) gives the maximum feedrate that can be used without producing an out-of-round hole, while Eq. (7.15) gives the time required to round up the hole and bring the part to "size." If a feedrate greater than $\bar{v}_{fopt}$ had been used and followed by a sparkout, a residual out-of-roundness would result.

In the grinding of cams, as shown in Fig. 7.10, the shape or contour ground into the workpiece often deviates slightly from the contour of the master cam. The cause of those errors in copying the master into the work blank is often related to the varying elastic deflections of the wheelwork system under the action of: (1) The grinding forces during the rough and finish portions of the cycle; and, (2) The threshold forces remaining in the system after the sparkout process.

During the grinding of cams, especially internal cams, variations in conformity, variations in the speed of the wheelwork contact zone, or "footprint," and variations in the threshold force occur as the wheel dives into the valleys and climbs to the crests of the cam profile.

The unique feature of internal cam grinding is that the threshold force varies from point to point around the contour of the cam. The threshold force, in turn, depends upon the instantaneous difference in curvature

Fig. 7.10 Internal cam grinding illustrating variation of conformity, footprint speed, threshold force, local wheel depth of cut.

$\dfrac{DW}{2}$ = Radius of bottom of cam lobe; $\dfrac{DS}{2}$ = Wheel radius;

U-U-U = Locus of used wheel (small); N-N-N = Locus of new wheel (large); ϕ = Angular displacement of work relative to wheel center; θ = Angular displacement of footprint

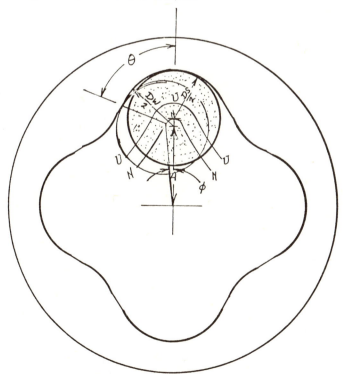

Δ, between the wheel and work, as measured by the "equivalent diameter" D_e.

Fig. 7.10 illustrates a wheel engaging the lobe of an internal cam. It will be seen that the D_e is very large when the wheel is in the concave region of the cam and very small when in the convex region. That causes a significant change in the threshold force intensity, from a relatively large value in the concave region to a small or negligible value in the convex region. Even though long sparkout times are provided, the residual deflection in the concave region cannot be eliminated. Accordingly, a contour error results. The magnitude of that error can be calculated from (Eq. 7.5) Contour errors may also be caused by variations in the speed of the wheelwork contact zone or "footprint." The angular position θ of the "footprint" in the concave zone is given by:

$$\frac{2 \sin \phi}{D_w - D_s} = \frac{\sin(\theta - \phi)}{A} \qquad (7.16)$$

Differentiating Eq. (7.16) with respect to time and setting $\frac{d\phi}{dt} = \omega$, the work angular velocity, gives the speed of the footprint:

$$\frac{d\theta}{dt} = \frac{2A \, \omega \cos \phi}{(D_w - D_s)\cos(\theta - \phi)} \qquad (7.17)$$

From that it will be seen that the "footprint" speed $\frac{d\theta}{dt}$ varies with angular position ϕ and may be much larger than the workspeed ω in the concave region, while it becomes less than ω in the convex region.

Since the wheel depth-of-cut h varies inversely with work surface speed according to:

$$h = \frac{\Lambda_w(F'_n - F'_{th})}{V_w} \qquad (7.18)$$

where: Λ_w = Work Removal Parameter

V_w = Work Surface Speed

the wheel depth-of-cut will be small in the concave region and large in the convex region. Consequently, a form error is ground into the workpiece during the rough grind. That error may never be completely eliminated during the sparkout if threshold forces are present.

Another source of form error is due to changing wheel size. Fig. 7.10 shows the locus of the center of the "New" wheel (curve NNN) and the locus of the center of a "Used" wheel (curve UUU). Notice that those loci differ considerably. Therefore, a cam grinder, with a fixed cross-slide motion governed by a master cam and follower, can only grind a correct workpiece for 1 wheel size. As the wheel wears smaller, form errors will again be produced.

Threshold force intensity is one of the important grinding machinability parameters that cause some materials to be very difficult to grind, difficult-to-obtain precision-size tolerances, consistent surface finish, roundness, and freedom from form errors as well as fast cycle times. The physical factors causing force thresholds have not been thoroughly studied and are not well understood. Those thresholds appear to be partly related to the resistance of a material to form a chip under the action of the abrasive grit and partly to hydrodynamic effects of the cutting fluid between wheel and work.

The Sparkout Process

After the grinding wheel has engaged the "black" workpiece and has rounded up the initial runout and removed the majority of the stock, it is necessary to bring the workpiece to the required size and taper tolerance, the required surface finish, microprofile, and surface integrity and to do those things in the least possible time. At the point C (Fig. 7.2), the cross slide stops and sparkout begins. The velocity of the grinding wheel $\bar{v}_w$ gradually begins to decrease as the spindle deflection and grinding force F_n start to decay. The grinding force decays down to the ploughing-cutting transition through the cutting region (Fig. 1.05), with the "cutting" metal removal parameter Λ_{wc}, or WRP, for the cutting regime. When the force intensity drops to the ploughing-cutting transition F_{pc}', the decay rate of the force becomes slower, corresponding to the smaller "ploughing" metal removal parameter Λ_{wp}. Further decay of the force intensity reaches the threshold force intensity F_{th}'.

At this point material removal ceases, and the wheel simply rubs on the work. A typical normalized grinding force intensity during a sparkout is shown plotted on semilog paper in Fig. 7.11. It will be seen that a kinked straight line results, corresponding to the 2 values Λ_{wc} and Λ_{wp}. Fig. 7.12 shows the instantaneous stock-removal rate during a sparkout[6].

It will be appreciated that the rounding process is most rapid when Λ_{wc} is large. Therefore, it is very desirable to have rounded up the workpiece before the wheel enters the ploughing regime.

It has been shown that the sparkout process follows an exponential

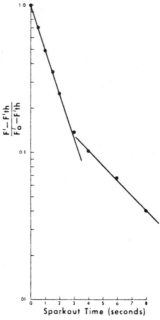

Fig. 7.11   Decay of reduced force intensity during sparkout   Wheel: 80K4   Dress lead—.004 in./rev.   Dress depth—.0005 in.   Wheelspeed—14200 FPM   Workspeed—250 FPM Coolant—cimperial 20   Work material—AISI 52100 R_c55

Fig. 7.12   Instantaneous stock-removal rate vs. normal force. Intensity during sparkout in Fig. 7.11

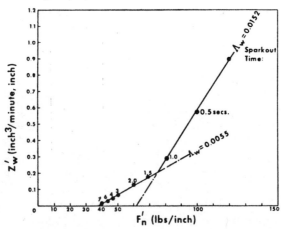

decay with a time constant τ_0[5] given by Eq. (7.8). Thus, 2 time constants occur corresponding to Λ_{wc} and Λ_{wp}, where distinct ploughing and cutting regions exist.

On grinding cycles where size, taper, and surface finish are determined by a sparkout of fixed time, several factors affect the size, taper, and finish variations. First, for cycles with a long sparkout time, the influence of the initial condition or force level from which sparkout began is completely lost. Size variations under those conditions are determined solely by variations in the threshold force F_{th}. For shorter sparkout times errors in size may be caused by variations in the initial force level, variations in Λ_{wc} and Λ_{wp}, and variations in the threshold force. For the case where the ploughing regime is negligible, a single sparkout time constant results and the quill deflection during a sparkout can be plotted as illustrated in Fig. 7.13. The solid line represents a standard cycle or, for instance, grinding with a new wheel that has been freshly dressed with a sharp diamond. The dotted line represents the situation where: (1) the initial force level has changed from F_n to $(F_n + \Delta F_n)$, (2) the time constant has changed from τ_0 to τ_1 because of changes in Λ_w in Eq. (7.8), and (3) the threshold force or corresponding quill deflection has changed by ΔF_{th}. Variations of those 3 factors can cause size, taper, and surface finish errors.

The influence of system rigidity K_t on size-holding ability at the end of a 6-second sparkout (about 3 time constants) is illustrated in the following table in the grinding of the 2 parallel ball tracks (Fig. 7.14) in automotive water-pump bearings[2]. In one setup the rigidity of the grinding quill and wheelhead was increased from 23,000 lb/in. to 50,000 lb/in. (4180 N/mm to 9100 N/mm) at the outer wheel—all other factors remaining the same.

It is seen that the more rigid system gives greater precision. That is primarily due to reduction of the variation in threshold quill deflection.

Variation of the threshold force F_{th} and the cutting and ploughing metal-removal parameters Λ_{wc}, Λ_{wp} as the grinding wheel becomes smaller can cause a disruption in size at wheel change in production internal grinders. That is illustrated by the size plot of the ball tracks in the double-groove automotive water pump bearing shown in Fig. 7.14a.

Table 7—1

System Rigidity lb./in.	Size-Holding Ability Std. Deviation σ(in.)
23,000	$110. \times 10^{-6}$
50,000	$32. \times 10^{-6}$

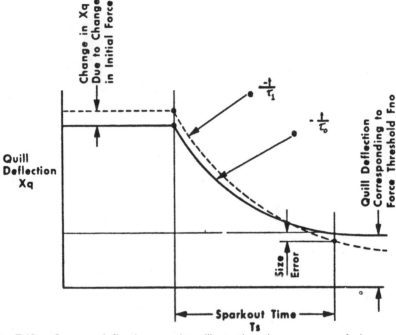

Fig. 7.13　System deflection vs. time illustrating three causes of size errors.

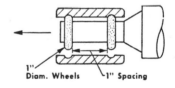

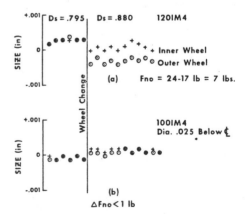

Fig. 7.14　Size plot of automotive waterpump bearing showing a wheel change for: (a) wheel spec's giving large threshold variation (b) wheel spec's giving small threshold variation

The bore size is shown plotted just before wheel change and just after wheel change. In Fig. 7.14a a size drop of about .0006 in. (0.15mm) and the appearance of .0004 in. (.01mm) size difference or taper between the 2 simultaneously ground ball tracks indicates higher force levels at the end of the sparkout for the new large wheel. That was confirmed by a measurement of the threshold force for the new and old wheels; 24 lbs (107N) for the new large wheel, 17 lbs (75N) for the used small wheel. That illustrates the use of "wheelwork characteristic charts," as shown in Fig. 1.05 (Chapter 1), for explaining and improving grinding performance. By selecting wheels that exhibit a very small change in F_{th} as wheel size changes, the size plot across a wheel change was improved to that shown in Fig. 7.14b, where a positive jump of about .0001 in. (.0025mm) occurred at wheel change.

On conventional grinding machines, the feedrate is controlled. As the grinding wheel engages the workpiece, forces are induced between wheel and work—the higher the force, the faster the stock removal. The induced force also governs the surface finish, the deflection in the machine, and the onset of thermal damage. Therefore, the induced force is one of the important variables that are uncontrolled in conventional feedrate grinding machines.

The ability of the cutting surface of the grinding wheel to remove stock, called the Wheel Sharpness and indicated by the WRP, is the second extremely important variable in the grinding process. In feedrate grinding, as the wheel sharpness drops (wheel becomes dull or glazed), the induced force rises (according to Eq. 7.15), resulting in increased deflection and, sometimes, thermal damage. Therefore, it is important to control the induced normal force and wheel sharpness.

Conclusions

From the previous sections it is clear that the normal force between wheel and work and the corresponding deflection of the system are key variables governing the size, roundness, and taper performance of many grinding operations. Methods for alleviating the effects of random stock variations, out-of-roundness, workpiece microstructure variations, and wheel sharpness variations are discussed in Chapter 14, "Adaptive Control in Grinding."

References

1.   David L. Brown, "Grinding Dynamics," Doctoral Thesis, Dept. of Mech. Eng., University of Cincinnati 1976, pp. 236.

2. R. S. Hahn, R. P. Lindsay, "Factors Affecting Precision in High-Production Internal Grinding", Mfg. Engineering Transactions v 2, 1973, SME, Dearborn, Mich.

3. R. S. Hahn, R. P. Lindsay, "On the Rounding-Up Process in High Production Internal Grinding Machines by Digital Computer Simulation," Proc. 12th Int. Machine Tool Design & Research Conf., Birmingham, England, Macmillan. 1972.

4. R. S. Hahn, "The Influence of Threshold Forces on Size, Roundness and Contour Errors in Precision Grinding," *Annals of C.I.R.P.*, 30, 1/1981, pp. 251–254.

5. R. S. Hahn, R. P. Lindsay, "The Influence of Process Variables on Metal Removal, Surface Integrity, Surface Finish and Vibration in Grinding," Proc. 10th Int. MTDR Conf., Univ. of Manchester, England, September 1969, pp. 95–117, Pergamon.

6. R. P. Lindsay, "Sparkout Behavior in Precision Grinding," SME Paper No. MR72-205, Dearborn, Mich. 1972.

CHAPTER 8

Centerless
Grinding

W. F. Jessup

Introduction

The external centerless grinder is a machine composed of an abrasive grinding wheel, a regulating wheel (usually abrasive), and a work support blade, usually surfaced with tungsten carbide. The purpose of the grinding wheel is to remove the metal and impart rotation to the workpiece. The function of the regulating wheel is to hold the part against the grinding wheel, cut pressure, and control the rotation. The purpose of the work support blade is to provide vertical support to the workpiece and to establish the proper support geometry necessary for producing part roundness (See Fig. 8.1).

Since on a centerless grinder the total support of the workpiece is obtained on its outside circumference, no work centers are required to establish an axis of rotation, hence the name "centerless."

There are 2 centerless processes: thrufeed and infeed. Throughfeed can be performed only on parts having a cylindrical surface of constant diameter along their lengths and with no protrusions that exceed the diameter being ground. On this operation, the parts follow 1 behind the other through the machine in an unbroken stream. Infeed (or plunge) grinding can be performed on parts having multiple diameters incorporating straight, tapered, or spherical surfaces. The parts can be ground 1 or several at a time spread across the wheels, which approach each other radially.

Extremely high degrees of geometric and surface finish accuracy can be obtained on both infeed and throughfeed operations (0.000008 in. for roundness and 1 to 2 microinches on surface finish).

The primary objective of a centerless grinder is to grind the circumferential surface of parts of rotation, the surface being a simple or complex profile generated about an axis. The purpose of that grinding operation is to refine the dimensional qualities of the parts and their surfaces to a higher degree of accuracy than that possible with the preceding operations. The accuracy is required for improving the mechanical characteristics of the part and/or for mating it with other components in the assembly for which it is intended. The 4 most common dimensions for which those accuracies are required are as follows: cylindrical straightness, or uniform diameter along the axis; axial profile, or dimensional relationships between different diameters along the axis; circularity, or dimensional uniformity from the axis to the surface being ground; and size, or constant diameter at 1 point along the axis relative to some absolute or discretionary reference (grinding to a fixed dimension or to match a mating part of varying dimensions).

Quality

Cylindrical Straightness

Cylindrical straightness is the part quality that assures that, with any axial displacement of the part diameter relative to its respective dimen-

Fig. 8.1   Basic elements of a centerless grinder: Grinding wheel, regulating wheel and blade.

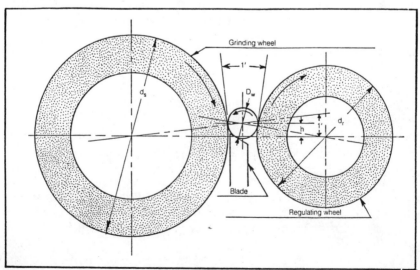

sion on the mating part, the clearance between the 2 surfaces will remain the same. On parts whose end function requires only rotary motion, good cylindrical straightness provides uniform radial support along the full effective length.

Axial Profile Accuracy

Axial profile accuracy is very important, because multiple diameter parts mate with a number of different components, each of which has its own manufacturing size dispersal spread. Each deviation from the tolerance for each diameter of the part must be deducted from the working tolerance, which should be available to ensure that there is proper matching of each diameter with its mating part. Axial profile errors throw a much greater responsibility on the sizing capabilities of the machine and, most of all, on the skill of the operator.

Circularity

Circularity is the most difficult of all the geometric-quality characteristics to maintain on a centerless grinder. The fact that the part is supported in a vee formed by the blade and the regulating wheel permits the generation around the circumference of a lobed profile. The general rules for setting up the machine to avoid making lobed parts are well-known, but the precise geometric dynamics that cause the setup to generate or correct the lobing are not yet fully understood.

There are several other influences in the process that can cause a part to be unround and that may or may not be related to the support geometry made up of the work contacts with the blade and regulating wheel and their relationship to the grinding wheel contact.

Size

Size, of course, is almost purely the responsibility of the machine and its equipment, except in the case of nonrigid parts, which are more or less independent of the positioning accuracy of the machine elements. There are 2 machine areas that have a nearly equal influence on the ability of the machine to repeat sizing positioning: the accuracy of the infeeding system and the static rigidity of all the machine elements in the force chain that holds the workpiece in contact with the cutting surface of the grinding wheel. They include the bed, the slides, and both spindles with their bearings. Another important problem in sizing is the thermal distortion of the machine and/or workpiece.

There is, however, a machine attachment that minimizes the negative effect of a low static stiffness in any of those elements and of machine thermals: size-control gaging. There has been a tremendous development in the technology of size-control gaging over the last 30 years, and it is possible now not only to gage-control the machine positioning but also to control most of the other variables in the process that detrimentally affect sizing. Since practically all of today's gages are electronically operated, the evolution of gaging technology will undoubtedly follow the curve of all other electronic development, and the whole problem of repeating the same size on consecutively ground parts will become one of the easier ones to solve.

Productivity

Industry requires not only that the parts be ground within the pre-established tolerances, but also that they be ground to those tolerances in the shortest possible time. The time required to produce a part on a centerless grinder can be divided between part-handling time, machine prepositioning time, and the time necessary to remove the metal while holding a given dimensional or surface quality.

Part-Handling Time

The centerless grinder is certainly 1 of the easiest of all metal-cutting machines to load and to unload after completing the operation. Before contact with the grinding wheel, the work lies freely in a simple vee. It can be injected and ejected axially, or it can be lowered radially from the top and unloaded radially down between the blade and regulating wheel. Or any combination of those 2 methods can be used. Of course, it must be realized that the configuration of the parts that can be gound on a centerless grinder lends itself to easy handling as compared with the oddly shaped parts that the centertype grinders are called upon to grind.

Prepositioning Time

The ease with which parts can be loaded into a centerless grinder makes it possible to load them with a minimum of clearance between the work and the wheel. Therefore, the prepositioning time of the machine can be kept to a minimum.

Metal Removal Time

Metal removal is the justification for the entire machine and concept. Irrespective of the results that are sought, they will be found only in removing metal. The time required to remove the metal will be a function of the amount that is to be removed and the rate at which wheel life and part quality will permit it to be removed.

The amount of metal that must be removed is a function of the depth of the imperfections in the surface of the preground part. Those imperfections can be mechanical, dimensional, or chemical. Or they can be any combination of the three.

Mechanical Imperfections

A mechanical imperfection could be a tool mark or a drawing crack that would weaken the mechanical resistance of the part if it were not completely ground out.

Dimensional Imperfections

Dimensional imperfections are the ones most commonly met. They are the result of the limitations of prior operations in holding tolerances that are required in the finished part. They are the bow that exists in all long, thin parts and hardened parts of nearly any length-to-diameter ratio. They are the ovality that is found in all bearing races. They are the eccentricity that can exist between any 2 diameters on a multi-diameter part. And there are certainly more.

Grinding from the Solid.   One major source of metal removal, but one that cannot be considered as an imperfection is the depth of profile of a formed part being ground from the solid.

Chemical Imperfections.   The 1 most common chemical imperfection that is found is the decarburized layer that exists on all parts that have been exposed to high temperatures in an oxygen atmosphere before grinding, either in a heat treatment or in a hot rolling or hot drawing operation.

Quality Requirements.   The quality requirements that must be respected after removing this metal have as great an influence on the grinding time as the amount that is to be removed. Nowhere does the classical compromise of "time vs. quality" apply more than it does in grinding. For any given state of the art, when quality requirements go up, the time required to produce them goes up by a similar order of magnitude. Therefore, for reasons of economy, it is extremely important that quality standards not be set any higher than the part function requires. If the equation "quality divided by time" is to result in a higher performance value, only an improvement in the state of the art can guarantee it. That means more efficient methods or, better still, new technology.

New technology is, of course, our best hope for any great improvement in the process efficiency because it is open-ended, whereas if we try to better the efficiency within the existing technology, the possibilities are necessarily limited. But new technology requires a better understanding of the basic process.

Basic Principles

To start an understanding of the process, it is first necessary to understand the basic principles of how a centerless grinder makes a cylindrical part. The best way to do that is probably to find the analogy with how a centertype grinder accomplishes a similar result because the same principles exist in both processes but they are easier to see and understand on a centertype grinder.

Principle of a Lathe

However little exposure an individual has had to metal-cutting processes, the 1 machine principle he or she probably does understand, because it is even more basic than a center-type grinder, is that of a lathe (Fig. 8.2). A lathe consists of 4 basic elements:

Fig. 8.2   The three principle cylindrical machining methods.

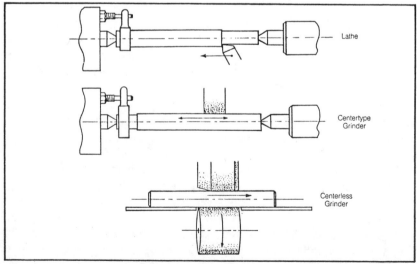

The Machine Centers.   There are 2 of them, each of which engages one of the 2 center holes that must be machined in the ends of the workpiece. The purpose of those machine centers is to provide radial support and axial location.

The Headstock.   The function of the headstock is to provide the part rotation that will permit the machine to accomplish its purpose of making a part whose machined surface is generated about an axis.

The Tool.   The tool is the metal-removing element or the element on which the entire process is based.

The Feed Mechanism.   For the tool to remove metal, an interference situation must be introduced between the tool and the workpiece. That interference is created by the feed mechanism's moving the tool in the direction of the workpiece until contact is made. The displacement can be along either 1 of the 2 axes of movement, radial or axial, that the feed mechanism provides to the tool. Of course, for metal to be removed as the result of that interference, it is necessary that the part turn. When the tool is performing circumferential machining operations, the radial axis provides the positioning function, while the axial movement and the part rotation cause the tool to cut. When face turning operations are being performed, the axial movement is used for positioning and the radial movement does the cutting.

Principles of a Center-Type Grinder

The principles of a center-type grinder are practically identical to those of the lathe, with 1 very important exception. The headstock rotation does not, in and of itself, cause the part to be machined. Its sole function is to present the work suface to the grinding wheel in a proper manner to permit the grinding wheel to remove the metal and to generate the required cylindrical surface. For example, contrary to the lathe, even if the headstock is stopped when the interference situation is created, metal will still be removed.

Infeed Grinding.   The radial support of the workpiece is obtained on a center type precisely as it is on a lathe: between centers. (Thus, the designation "center type.") On a lathe the cutting edge of the tool is relatively narrow; consequently, the radial feed is used only occasionally for operations other than facing an end or a shoulder. On a center-type grinder, where the cutting portion of the tool can be quite wide, grinding over the total axial length of the part to be ground can be done using the radial feed. That is called "infeed" or "plunge" grinding and is possible on any part whose ground length is less than the working width of the grinding wheel.

Traverse Grinding.   There are, however, many parts whose ground length is greater than the working face of the grinding wheel. On those parts it is general practice to use a medium-width wheel and traverse the workpiece axially while maintaining the interference contact of the wheel with the work. That is called "traverse grinding." However, since only a small interference penetration is permitted (in order to spread the cutting contact over the greatest possible width of the wheel), that penetration usually represents only a small part of the total depth of stock material to be removed. Consequently, many traverse strokes are required to arrive at the final diameter. That permits the part to be loaded into the machine once, and, irrespective of the number of passes required to finish the operation, it stays in the machine until the operation is completed.

Principles of a Centerless Grinder

It is possible to show that all of the elements that are present in a center-type grinder and that are necessary for grinding a cylindrical part are also present in a centerless grinder, even though it may be difficult to recognize their functional relationships.

There are the same 4 basic process elements: support, rotation, tool, and feed, but on the centerless, several of them can be combined in the same machine element. For example, in centerless plunge grinding, the radial support is provided by the blade and the regulating wheel, but the regulating wheel also ensures the rotation. On a center type, the feed is obtained by a feed mechanism that displaces radially either the wheel or the work to create the interference condition that is necessary for metal removal. In throughfeed grinding (the equivalent of traverse grinding on a center type) the regulating wheel performs 3 of the 4 basic functions in the process: support (shared by the blade), rotation, and axial feed. That leaves only 1, but the most important for the grinding wheel: metal removal.

Regulating Wheel Functions.   By far the most important machine element in a centerless grinder that differentiates it from a center-type grinder is the regulating wheel. On a center-type grinder, the rotation is imparted to the workpiece by the headstock, positively driving it through a rotating driver acting circumferentially on a dog that has to be clamped to the workpiece. On a centerless, the rotation is imparted by the grinding wheel (the work rotates with the grinding wheel, contrary to a centertype), and the regulating wheel acts only as a brake to control that rotation. (The regulating wheel is also frequently called the "control" wheel. They are both appropriate names that truly describe its role in providing the required workspeed. It is also called occasionally the "drive" wheel, which is a misnomer in terms of its real function.) Since the regulating wheel exerts no positive rotational force on the workpiece, its control over the speed comes purely from the friction that exists between work and regulating wheel. There are other factors involved in workspeed control, but regulating wheel friction is the most important. (Some others are blade angle, blade friction, cutting fluid lubricity, and the ratios of the diameters of the workpiece).

Two of the most important advantages of the centerless over the center type is that the workpiece does not require that a driving dog be clamped to it, and no centers have to be machined in the ends of the workpiece.

The feed function of the regulating wheel in throughfeed grinding is derived from the tilting of the axis of the regulating wheel in a plane that is parallel to the surface of the regulating wheel at the point of contact with the work. For example, on the most common machine configuration—having the 2 wheel axes lying on the same horizontal plane—the plane of the feed tilt is vertical. On the less common type, where the grinding wheel lies directly above the regulating wheel, the tilt angle is in a horizontal plane.

As the regulating wheel rotates the work, the angle of the regulating wheel axis to the work axis imparts an axial feed motion to the work, whose rate is the product of the regulating wheel diameter, the regulating wheel rpm, and the sine of the real feed angle. (Since feed angle graduations are taken from the horizontal for the zero base, any tilt of the work axis in the vertical plane—for example, from a badly aligned blade—will modify the angle from that which is indicated on the graduations).

A hypothetical image that may help to understand how the thrufeed works is that a point on the regulating wheel surface picks up the equivalent point on the workpiece surface at the angle of tilt and then drops it back. The new position of the point on the workpiece will be at a new axial position from where it was picked up. That new axial position will be farther advanced from the original position by an amount equal to the distance it was picked up before dropping, multiplied by the sine of the feed angle. If that operation were repeated, with the point on the work being again picked up by a new point on the wheel, carried forward, and dropped, it would have fed one step farther forward. If the process were to continue to repeat, the feed would be intermittent, but the part would advance between the wheels. If we now make the lifting height infinitely small, it could be considered to be a continuous throughfeed and it would describe precisely the throughfeed that is actually obtained.

On center-type grinding of workpieces that are longer than the width of the grinding wheel, the workpiece is traversed back and forth past the cutting face of the grinding wheel, a small part of the total stock being removed with each pass, until the part is reduced to size. When one is throughfeed grinding on a centerless grinder, only one direction of axial feed is possible, since the angle that provides that feed can be set in only 1 direction. Consequently, only 1 passage through the machine is possible for each pass of metal removal.

There are, however, several unique characteristics of the centerless grinder that easily compensate for whatever disadvantages might result from having the part pass through the machine several times. First of all, since the workpiece is supported over its entire grinding length, much wider wheels can be used than on a center type (where the part is supported only at the extreme ends or, occasionally, by a limited number of additional supports in the middle in the form of steady rests). Those same wider wheels can also remove more metal per minute per unit of wheel width because of this support. That means that many fewer passes are required to remove the total stock. And because the feed is always in the same direction, it is possible to set an angle between the grinding wheel and the regulating wheel, open to the incoming side, to distribute the stock removal uniformly over the entire cutting surface of the grinding wheel. That uniform stock removal distribution is the most important factor in ensuring the maximum grinding-wheel wear efficiency.

Center-type grinding is a batch-type operation, whereas throughfeed centerless grinding is a continuous process. What that means is that while one is loading a center type, no grinding can be done and thus loading becomes a negative value in the efficiency calculation. In throughfeed centerless grinding the loading operation is performed simultaneously with the grinding operation; consequently, it subtracts nothing from the efficiency of the operation.

There is, however, 1 complication introduced into the process by setting a feed angle on the regulating wheel. As was said previously, the regulating wheel serves several functions. Both the feed function and the rotation function have been described previously, but the support function is tightly tied in with the feed function.

In the support function, what is absolutely necessary is *complete* support. Since the major justification of the machine lies in the part's being supported over its full grinding length from which is derived the high manufacturing efficiency of the process, a natural assumption is made that the resistance to the grinding cut will be there, and that sufficient mechanical stability will always be available to guarantee the required geometric quality. That requires, however, that the support must really be there over the whole grinding length. If it is not, the part will rock or it will bend or it will chatter or all 3. What is guaranteed is that if the full support is not there, the quality will suffer.

When 2 cylinders, workpiece and regulating wheel, are brought into contact, with their axes perfectly parallel, the contact between the 2 is an unbroken line from end to end. That is what the contact would be between the regulating wheel and a cylindrical workpiece if the regulating wheel were dressed to a perfect cylinder and the throughfeed angle set at zero. If the grinding wheel is also dressed to a perfect cylinder, so that the workpiece remains cylindrical throughout the grind, the contact between the regulating wheel and the workpiece will remain a straight, unbroken line, providing good support and guaranteeing good geometric quality of the finished part.

If, however, the workpiece axis is maintained in a horizontal line while the regulating wheel is tilted to some feed angle in the vertical plane, the contact between the work circumference and the regulating wheel circumference will no longer be a straight, unbroken line but will become a point in the middle of the workpiece. Consequently, the part will be completely unsupported at the ends, and it will be chattered and/or out of round. The only possible solution for that problem is to change the shape of the regulating wheel to bring the support to the work at the unsupported ends.

A theoretically ideal method of accomplishing that would be to replace the workpiece in its grinding position against the regulating wheel by a rotating diamond tool that is the exact size and shape of the workpiece. As the regulating wheel is turning, the diamond roll will dress away that portion of the regulating wheel circumference that had presented a heavy contact of the work on the wheel. Eventually, the entire surface of the diamond roll would be in contact with the regulating wheel, and when it was replaced again with the workpiece, there would be full contact between work and regulating wheel over the entire grinding length of the part.

In the beginning years of centerless grinding, around the early 1920s, the regulating wheels were actually dressed by a method very similar to this, except that instead of a diamond roll being removed (which had not even been thought of at that time), the blade was removed and replaced with a dressing device that passed a single-point diamond along a path that followed the theoretical contact line of the work with the regulating wheel. That contact line was sufficiently accurate to permit parts to be ground within the tolerances of the time, but tearing down the blade setup, removing the dressing device, and replacing the blade setup was an onerous task and prohibitively time-consuming. Another way had to be found. (Actually, today, appoximately half of the centerless grinder manufacturers in Europe use that basic concept, with only minor variations. Instead of passing the diamond across the wheel at the front, in the workpiece position, they pass it across the wheel at the rear, opposite the workpiece position).

The method chosen in the United States to generate the proper geometry into the regulating wheel was to divorce the dressing geometry from the feed function completely. To do so, the truing device was repositioned approximately 100° around the regulating wheel toward the top. That required a separate angular adjustment of the diamond displacement in a plane parallel to the tangent to the wheel at the point of contact with the diamond. That angular adjustment is designed to permit the diamond to follow a path that has the same geometric relationship to the wheel as the work contact has in grinding position.

With that concept of a separate truing device for obtaining the proper axial regulating wheel profile, the resuts in grinding were quite the same as truing in grinding position, but without the time lost in changing over from grinding to truing. However, in most truing methods it was observed that the workpiece was frequently low in the middle, particularly on parts whose length approached the width of the wheels. By juggling

Fig. 8.3   Truing and feeding from the same tilt angle.

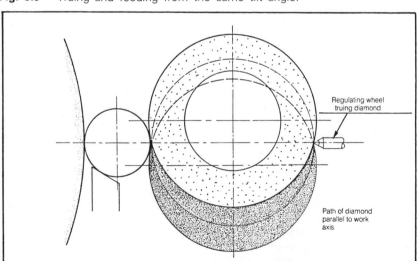

Regulating wheel
truing diamond

Path of diamond
parallel to work
axis

the feed angle adjustment, it was found possible to arrive at a setup that produced cylindrical parts. After many such experiences, it was discovered that to make cylindrically straight parts, there had to be a slight difference between the feed angle set on the regulating wheel and the corresponding angle set on the truing device. That difference is the result of the fact that the wheel is trued with a point and the work contact is an arc (see Fig. 8.3).

The difference in angles required is a function of work diameter, regulating wheel diameter and feed angle and, as mentioned, is based on the assumption that the diamond is a point. Since we know the geometric base of the phenomenon, it has been possible to derive an equation for calculating the angle difference. The equation is:

$$\alpha_1 = \alpha/\sqrt{1+D_w/D_r} \tag{8.1}$$

Where:

$\alpha_1 =$ The truing angle
$\alpha\ =$ The feed angle
$D_w =$ Workpiece diameter
$D_r\ =$ Regulating wheel diameter

There is still 1 more adjustment that is an essential part of the truing operation: "diamond offset"

The diametric size that is produced on a centerless grinder is a function of the distance between the cutting face of the grinding wheel and the support face of the regulating wheel at the point of contact of the workpiece. The circumferential positions of those points of contact are maintained throughout the grinding operation by the support blade. During the grinding operation the part rotates within the confines of this 3-sided figure, and since the distance between the diametrically opposite wheel contacts is, for explanation purposes, constant, the machine will grind a part of constant diameter when measured at any position along the part length. By most standard methods of measuring cylindrical parts, "constant diameter" would be interpreted as being "round," but on a centerless grinder, that is not necessarily the case. The centerless setup geometry (work contact with the regulating wheel, blade, and grinding wheel) permits grinding parts with constant diameter when measured diametrically, but when measured in a vee-block they can show large errors of circularity.

To produce a part that will measure round when measured in a vee-block as well as diametrically, it is necessary that the blade height be set so that the axis of the workpiece lies above the centerline that passes through the 2 wheel axes. (A detailed description of that phenomenon will be given in a later section). Because the part lies above the centerline, the intersection between the 2 tangents drawn between the contact points of the workpiece and the 2 wheels forms an angle that defines the roundness correction geometry of the centerless grinder. That angle is designated by the Greek letter γ, or gamma. Since a gamma angle is absolutely essential for making round parts on a centerless grinder and making round parts is the primary objective of the process, it must be assumed that there will always be a gamma angle (see Fig. 8.1).

If a workpiece is lying on the blade with its axis on the wheel centerline and bearing on a regulating wheel that has been dressed with a feed angle that produces a symmetrical form (the 2 ends of the wheel the same diameter), the axis of the work will be parallel to the axis of the grinding wheel. If, however, we now raise the work axis above the wheel centerline, the front edge of the regulating wheel will push the part into the grinding wheel and the back edge will permit the part to fall back away from the grinding wheel. Consequently, the axis of the part will no longer be parallel to the axis of the grinding wheel and the contact between the 2 will be heavy at the front.

In order to dress the regulating wheel to a form that will reestablish the parallelism between the workpiece and the grinding wheel axes, it is necessary to displace the diamond off the center of the regulating wheel by an amount equal to that which the height of the contact line between work and regulating wheel is set above the centerline of the wheels (see Fig. 8.4).

Fig. 8.4   Setting the diamond offset.

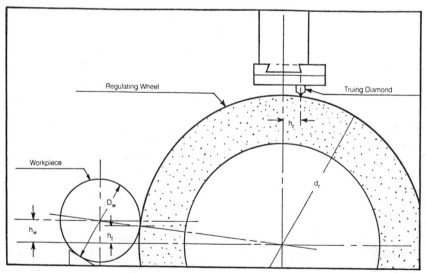

The formula for setting this diamond offset is:

$$h_t = h_w D_r / (D_r + D_w) \qquad\qquad (8.2)$$

Where:

h_t = Truing height
h_w = Work center height
D_r = Regulating wheel diameter
D_w = Workpiece diameter

Geometric Quality

There are 5 main categories of geometric quality errors:

1. Circularity (Roundness)
2. Axial profile accuracy (Step relation)
3. Cylindrical Straightness
4. Concentricity
5. Bow

Circularity

This is the category of geometric quality defects that is thought of as being the one most commonly associated with the centerless grinding process and with good reason. Because of the rather poorly defined axis of rotation that is inherent in the process, there are several sources of roundness errors that can creep into the operation. The most important is from the centerless setup geometry.

Centerless Setup Geometry.   The basic reason for the tendency of a centerless grinder to generate a lobed diameter on a part lies in the centerless setup geometry, which is composed of the contact between the work and the blade, the regulating wheel and the grinding wheel. That configuration of work, wheels, and blade belongs to a family of geometric figures involving 2 polygons, 1 of which, the inner polygon, can rotate within the confines of the other polygon, the framing figure, and maintain contact at all times with the sides of the outer polygon. (Although the surface of the wheels that make up the 2 opposite sides of the framing figure are arcs, the radii of curvature are so large compared with the relatively small errors on the workpiece that they can be considered as straight lines) (see Fig. 8.5).

Since the centerless geometry of a machine set up with the axis of the

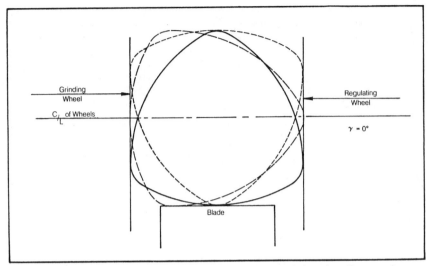

Fig. 8.5   The wheels and blade define a four-sided framing figure when grinding on center.

part lying on the centerline of the wheels forms a square, with the 2 sides representing the wheels being parallel, any inner polygon having an odd number of sides will "fit" (permit rotation without losing contact with any of the four sides) (see Fig. 8.6).

The inner polygons have a clearly defined form, called a "polycentroid," having pairs of radii around a number of centers equal to the number of lobes in the polygon (see Fig. 8.7). With that figure the diameter of the polycentroid is constant around the circumference, being the sum of the 2 different radii. In order to destroy "fit," we must change the parallelism of the sides of the framing figure by lifting the part center above the centerline of the wheels. By doing so, we introduce a gamma angle.

If we now rotate our workpiece, represented by the polycentroid, in the grinding position, represented by the framing figure with vertical sides that are no longer parallel, we will see that an interference is created between the small radius of the polycentroid and the grinding wheel. Consequently, as the workpiece is rotated and the wheels are brought closer together, with each revolution of the workpiece the common centers of the radius pairs will approach the axis of the workpiece and the deviation from a true circle will diminish (see Fig. 8.8).

A highly simplified explanation of what happens is that as the contact point of the rotating polycentroid slides along the reference side of the framing figure (regulating wheel side), a point diametrically opposite will always move parallel to the reference side and vertically along the

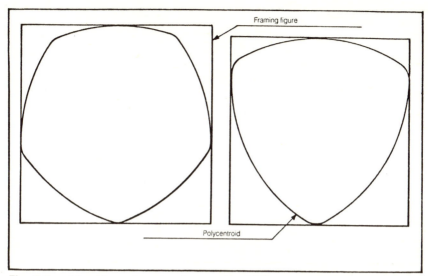

Fig. 8.6   All odd numbered polycentroids "fit" a square framing figure.

Fig. 8.7   Construction of a polycentroid.

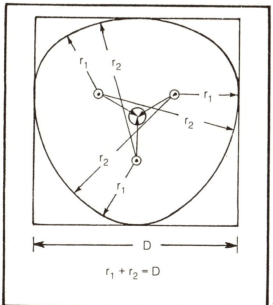

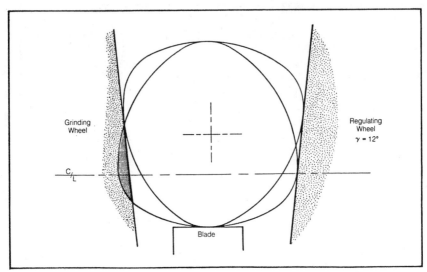

Fig. 8.8   Correction interference created by a 12° gamma angle.

contact face of the grinding wheel. But since that wheel is not parallel to
the reference side and is closer to the reference side the nearer it is to the
wheel centerline, it is at that point that the small radius of the polycentroid is in contact with the grinding wheel. Consequently, more material is removed from the small radius than from the large radius, which
is in contact with the grinding wheel only at a point higher above center
where the wheels are farther apart.

A common point of some confusion in setting up a centerless grinder
is whether or not it is possible to grind below the centers. (The axis of
the workpiece would be displaced below the plane connecting the two
spindle axes. That would be called a negative gamma angle).

If we return to the principles of the "framing figure/polycentroid"
couple, we will recall that on an odd-numbered polycentroid the periphery opposite a straight reference line will always slide along an imaginary line parallel to the reference line. In the case where the blade is the
reference line, there is an imaginary line parallel to it (and at a distance
from it equal to the diameter of the workpiece) along the top of the part.
(See Fig. 8.9.) If we now invert the geometry by 180° and replace the
imaginary line with the line of a blade and consider that what was the
blade will now become the imaginary line, we will have exactly dupli-
cated the geometry that is necessary for the rounding-up effect of the
workpiece (see Fig. 8.10). Therefore, there should, theoretically, be no
difference in the rate of roundness correction obtained on a centerless
setup using a flat-top blade with the axis of the workpiece offset from

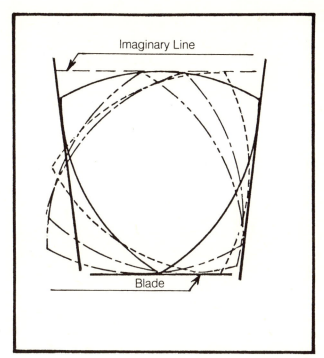

Fig. 8.9 Grinding above center on a flat top blade

Imaginary Line

Blade

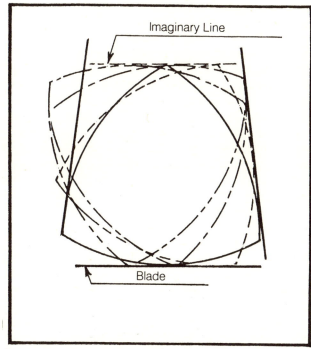

Imaginary Line

Blade

Fig. 8.10 Grinding below center on a flat top blade

the centerline either positively or negatively. In practice, however, there is some difference, with the positive gamma angle producing somewhat better roundness correction than a negative gamma angle. Another difference between the two setups is the direction of the vertical component of the cut force which, when grinding above center, reduces the pressure on the blade, and, when grinding below center, increases the pressure against the blade. Although that change has some slight effect on the machine structure, it is primarily significant in its effect on the friction between workpiece and blade and in its influence on mechanical stability of the workpiece support.

In the vast majority of centerless applications, an angle top blade is used in preference to a flat top blade. Experience has proved that the rate of roundness generation per revolution of the workpiece is increased for a given gamma angle by introducing an angle to the support surface of the blade. A simple graphic sketch of the wheelwork interference pattern during rotation of the workpiece in a setup composed of a positive gamma angle and a positive blade angle shows quite readily that more material is removed from the small radius relative to that removed from the large radius than is true in a setup having a positive gamma angle and a flat top blade. (See Fig. 8.11.) This beneficial effect of the blade angle results directly from an increase of vertical movement of the axis of the workpiece, thus enhancing the difference in interference between the grinding wheel and the large radius and the grinding wheel and the small radius.

Since the rate of roundness generation of the workpiece is a function of the vertical displacement of its axis during the rotation, it may be a good idea to look at what happens with a certain polycentroid supported on a blade of a certain angle. A 5-lobed polycentroid riding on a blade with an 18° top angle will produce a vertical displacement of its axis that is less than if it were on a flat top blade. That means that the rate of roundness generation per revolution will not be enhanced by the addition of the 18° blade angle for correcting roundness on a 5-lobed part, though it will improve the roundness generation of a 3-lobed part. (See Fig. 8.12.) The result will be that the 3-lobed roundness will be corrected out of the part long before the 5-lobed roundness will be corrected.

Although the 30°-angle blade that is standard does a good job for correcting 3-lobed out-of-roundness, that blade does not do so well for correcting 5-lobed out-of-roundness. (See Fig. 8.13.) This is the reason that the final lobing that must be corrected out of a part ground on a 30°-angle blade is of 5 lobes when one is grinding to very close tolerances.

However, by going to a 45° blade angle, the roundness is improved for both 5- and 3-lobed parts, but the greatest improvement is in the 5-lobed

Fig. 8.11 Grinding above center on an angle top blade

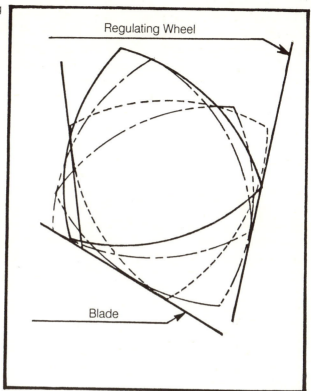

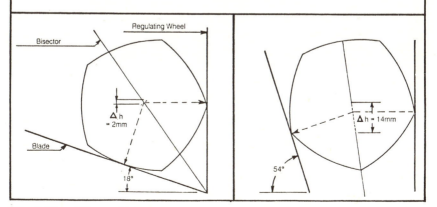

Fig. 8.12   Effect of blade angle on center height displacement during grinding.

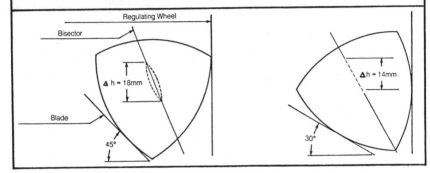

3—LOBE POLYCENTROID; RADIAL LOBING AMPLITUDE = 8MM

When a part is rotated in support contacts that are separated by full lobes the geometric center displaces along the bisector. When rotating on support contacts that are separated by half lobes the geometric center displaces normal to the bisector. (See Fig. 8-12) When the part is rotated in support contacts that are separated by other than exact multiples of half lobes the geometric center displacement will describe an eliptical path, with the major or minor axis falling on the bisector.

Fig. 8.13

out-of-roundness. Consequently, it is always recommended to use a 45° blade angle when you are grinding to very close tolerances for roundness. (See Fig. 8.14.)

If, however, an angle-top blade is applied to a setup incorporating a negative gamma angle (with the axis of the workpiece set below the centerline of the wheels) a totally different situation obtains. With that geometry, there is a reversed phase relation between wheelwork contact and the presentation of the 2 different radii to the wheel. It can be shown graphically. (See Fig. 8.15.) that maximum interference occurs at the time the large radius is in contact with the grinding wheel, and minimum interference occurs when the small radius is in contact with the grinding wheel. Consequently, the small radius becomes smaller and the large radius becomes larger, with the result that the centers of the radius pairs move away from the axis of the workpiece toward the circumference of the part, increasing its polygonal out-of-roundness.

Since the influence of the gamma angle on the entire centerless process is so important, through its positive and negative effect on the quality of the workpiece, it deserves to be singled out for a great deal more attention than it has ever received in the past. Since an increase in gamma increases the rate of roundness generation per revolution of the part, but at the same time it sensitizes the machine to self-excited chatter, there is always some compromise setting that will be optimum for a given set of conditions. That means that, theoretically, there is only 1

5—LOBE POLYCENTROID; RADIAL LOBING AMPLITUDE = 3MM

This comparison of centerless setups using blade angles of 30° and 45° shows that there is a greater vertical displacement of the geometric center of the part on the 45° angle blade than on the one with 30°, hence a better roundness correction per revolution of the part for a given gamma angle.

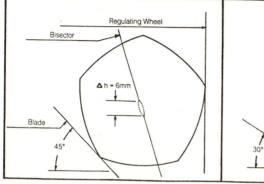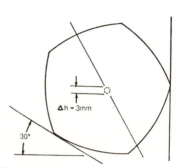

Fig. 8.14

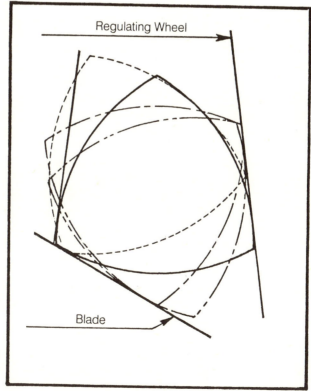

Fig. 8.15 Grinding below center on an angle top blade

correct gamma angle that will guarantee the best possible results. Actually, however, we know that there must be considerable latitude, since it has never been used in the past as a unique parameter and has been approximated only by working strictly with the height above center. The successful use of centerless grinders for the past 60 years is proof that good parts can be made by that approximation. However, with better understanding of the process coinciding with demands for better and better quality, a more accurate control over gamma can assure a better control over part quality.

The problem that has deterred centerless operators in the past from using gamma as a setup parameter is that it cannot be set by a single adjustment of a machine element, whereas a height above center can be established by the simple adjustment of the blade to a measurement from the axis of the workpiece to a reference surface on the machine. The problem is that the gamma angle not only is related to the height of the work above center but also is greatly influenced by the diameter of the grinding and regulating wheels and the diameter of the workpiece. In fact, a machine having a 24-inch diameter grinding wheel and a 14-inch regulating wheel, grinding a 2-inch diameter workpiece at 5/8 inch above center will start out with a gamma angle of 7.25° with new wheels. But with completely worn wheels on both sides, that angle will increase to 10°, representing an increase of more than 35%. What that means is that a machine that will grind properly with a gamma angle of 10° with no problem of self-excited chatter, and still have the attendant advantages of increased roundness correction, would certainly benefit from having the 10° gamma angle throughout the life of both wheels.

Of course, what that requires is that the operator would have available means of calculating the proper height above center for the proper gamma angle and also have means for easily setting that height on the blade. And on throughfeed operations, there would still remain the adjustment that would be required to produce the equivalent of a change in diamond offset that corresponds to the work height above center.

For the time being, the adjustments in the machine do not yet exist that will permit an "in-process" control for a constant gamma angle. Still, the other major variable in a gamma calculation—the work diameter—can be controlled so that only the wheel diameter variable will cause the gamma to deviate from optimum.

To start with, the simplest possible formula is necessary to make that calculation. A mathematical formula that is perfectly accurate mathematically but reduced to its simplest expression is as follows:

$$\gamma = \text{Arc sin} \left(2h_w \left(1 / (D_s + D_w) + 1 / (D_r + D_w) \right) \right) \qquad (8.3)$$

Where:

γ = Roundness correction angle (see Fig. 8.1)
h_w = Height above center
D_s = Grinding wheel diameter
D_r = Regulating wheel diameter
D_w = Workpiece diameter

That formula is used to monitor the gamma angle of a setup where the height above center is the starting variable.

To set the height above center to obtain a desired gamma, the following formula is used:

$$h_w = \sin \gamma \, / \, 2 \, (\, 1 \, / \, (D_s + D_w) + 1 \, / \, (D_r + D_w) \,) \qquad (8.4)$$

Based on the fact that the maximum gamma angle that will be used in the normal range of centerless applications will be approximately 12°, and throughout that range the sine tables deviate from a linear progression by only approximately 0.7%, and in practice it is impractical to set the work height to an accuracy even close to that, a numerical constant that is calculated as the inverse of sin 1° can be inserted in the formula to replace the sine function. Since that can be combined with the other numerical constant, 2, that appears in the above formula, we then arrive at a new formula with a numerical constant of 115, which is a more-than-adequate approximation of the above mathematically accurate equation and which is as follows:

$$\gamma = 115 \, h_w \, (\, 1 \, / \, (D_s + D_w) + 1 \, / \, (D_r + D_w) \,) \qquad (8.5)$$

Since that formula will be used only for monitoring the gamma angle of a setup that has been made according to a known height above center, a rearrangement of the parameters for setting the proper height to obtain a gamma required is as follows:

$$h_w = \gamma \, / \, 115 \, (\, 1 \, / \, (D_s + D_w) + 1 \, / \, (D_r + D_w) \,) \qquad (8.6)$$

The advantage of replacing the sine function with a linear constant is that the calculation can be made on any simple 4-function calculator.

Beta Ratio.   The beta ratio is the ratio of grinding wheelspeed (rpm) to workspeed (rpm). The origin of the roundness problems it creates is in the fact that all grinding-wheel spindles run out, perhaps only 5 or 10 millionths of an inch, but they *do* run out. If the grinding-wheel spindle runout is synchronized with the workpiece rpm in a ratio that makes a whole number, the same side of the grinding wheel will always contact the workpiece at the same circumferential position and the part will be out-of-round.

If it is an even-numbered integer in the beta ratio, the part out-of-roundness will be equal to the spindle runout and a roundness error of 5 to 10 millionths of an inch cannot be found in the average job tolerances. However, if the beta ratio is an odd-numbered integer, the error is added to the part with each revolution and the accumulation can quickly exceed most ordinary tolerances. When the beta ratio is a whole odd number, the out-of-roundess generated into the part will have a lobing frequency that is equal to the beta ratio. That is called a simple beta ratio. However, there is a whole family of beta fractions that will make beta polygons. A beta fraction is made whenever an odd-numbered integer is divided by any whole number that is smaller than the integer. (See Fig. 8.16.) The beta polygon made by a beta fraction will have a number of lobes equal to the odd-numbered numerator. They are called compound beta ratios.

Simple beta ratios are made when the high side of the wheel hits the workpiece at the low side of the polygon between each 2 lobes. Compound beta ratios are made when the high side of the wheel contracts the workpiece at a spacing around the circumference equal to a number of lobes that is in the denominator of the fraction. (See Fig. 8.17.)

Self-excited Chatter.   Self-excited chatter is a problem on machines that do not have a high dynamic stiffness. The lack of dynamic stiffness can be a weakness in the original machine design, or it can be from wear. The chatter marks produced in the surface of the work will nearly always be even-numbered and will display a smooth sinusoidal profile on a roundness-testing instrument.

Both height above center and workspeed can act as triggers for self-excited chatter. As the work is set higher above center, the machine becomes more sensitive to chatter, and as the work turns more rapidly, chatter sensitivity increases. Since those are the 2 adjustments that are used for increasing the rate-of-roundness generation, they must be set at optimum values: high enough to produce the maximum rate-of-roundness generation but low enough to avoid chatter.

Chatter having its origin in the machine structure is not the only chatter than can be had with a centerless grinder. The workpiece itself can induce chatter. Parts that have an unground portion on 1 end that extends out from the end of the workpiece can chatter. The chatter is usually of fairly high frequency and only at the end near the unsupported portion. On that chatter problem, workspeed is the critical variable. A threshold speed can be found above which the part will chatter and below which it will not chatter.

The blade can also be a source of self-excited chatter. If it is not clamped tightly in the workrest or if the workrest is not solidly fixed to the mounting surface or if the blade is too thin for the work load that is

BETA RATIO, "β"

If the ratio of grinding wheel rpm to workpiece rpm calculates out to be a fraction, "f", whose numerator is a whole odd number and whose denominator is any whole number less than the numerator, a polygon having a number of lobes equal to the numerator will be generated on the workpiece. The lower the number in the denominator, the greater will be the amplitude of the lobing.

Beta ratios that can produce odd-numbered polygons of "n" lobes are:

n=3		n=5		n=7	
f	β	f	β	f	β
3/2 = 1.5		5/4 = 1.25		7/6 = 1.17	
3/1 = 3		5/3 = 1.67		7/5 = 1.4	
		5/2 = 2.5		7/4 = 1.75	
		5/1 = 5		7/3 = 2.33	
				7/2 = 3.5	
				7/1 = 7	

The formula for calculating Beta is as follows:

$$\beta = n_s\, D_w / d_r\, n_r$$

Where:

β = Beta ratio
n_s = Grinding wheel rpm
D_w = Workpiece diameter
d_r = Regulating wheel diameter
n_r = Regulating wheel rpm

Fig. 8.16 Beta fractions that can make beta polygons.

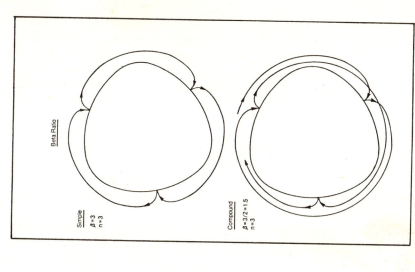

Beta Ratio

Simple
$\beta = 3$
$n = 3$

Compound
$\beta = 3/2 = 1.5$
$n = 3$

Fig. 8.17 Simple and compound beta ratios.

asked of it, chatter can result. The frequency is usually much higher than it would be for any part of the machine structure—in the range of 250 to 300 Hz.

Forced Vibrations.   Forced vibrations are those that are introduced into the grinding operation but come from a source outside the operation. They can come from the machine itself, or they can come from another machine or operation completely separate. The most common source of forced vibrations is from the grinding wheel, the grinding wheel motor, or the drive belts to the grinding-wheel spindle.

Chatter induced by forced vibrations does not show up on a roundness-measuring instrument as a clean, sharp frequency, but as a roughness in which it is sometimes difficult to define a frequency. It also affects the surface finish, adding a few microinches. That is not too important on normal tolerance operations but can be quite critical on very accurate operations.

Interrupted Surface.   Interrupted surface workpieces (with keyways, oil grooves, holes, etc.) will always have some out-of-roundness unless the interruption is a very small proportion of the length of the surface being ground. Such workpieces fall into 2 categories: relatively long parts ground 1 at a time, infeed or throughfeed, and short parts ground throughfeed.

The out-of-roundess ground into such workpieces is the result of deflection in the machine structure and/or in the regulating wheel support surface. On long parts ground 1 at a time, both the machine structure and the regulating wheel deflection enter into the problem. On short parts being ground throughfeed, only the regulating wheel has any influence on the part out of roundness because of the random positioning of the interruptions in the individual parts stacked between the wheels.

When the interruption is against the grinding wheel, there is less metal being removed than when the side opposite the interruption is against the grinding wheel. Less metal being removed means less normal force to push the part into the surface of the resilient regulating wheel. And at that same time, the full uninterrupted surface is against the regulating wheel, giving the maximum support. (See Fig. 8.18.) As the part rotates 180°, the interrupted side of the part is in contact with the regulating wheel and the uninterrupted side is in contact with the grinding wheel, which applies the maximum pressure on the part. That maximum pressure must be supported by the smaller support surface in contact with the regulating wheel, and the axis of the part shifts toward the regulating wheel, leaving metal on the uninterrupted side.

As the uninterrupted side turns again to the regulating wheel, the

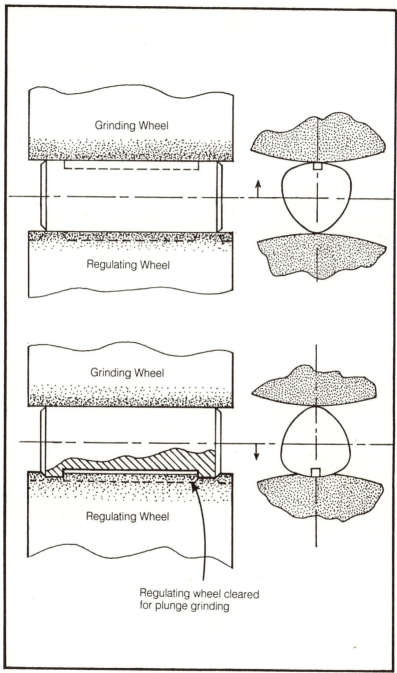

Fig. 8.18   Interrupted surface phenomenon.

metal that was left on the part pushes it into the grinding wheel, removing the same amount from the interrupted side plus the amount that is due to the axis's shifting back toward the grinding wheel.

On infeed operations it is possible to clear the bearing on the regulating wheel under the interruption and eliminate the full effect of the change of support length on the regulating wheel. It does not, however, change the difference in radial pressure on the part coming from the difference in length of contact with the grinding wheel. Consequently, it only diminishes the axis shift; it does not eliminate it.

On throughfeed operations, it is not possible to clear the bearing under the interrupted surface; consequently, the full effect of the change of support length on the regulating wheel must be accepted. There is, however, something that can be done to minimize the effect. Since the whole phenomenon is based on the deflection of the surface of the regulating wheel, it is necessary only to use a very hard regulating wheel.

The rate-of-roundness correction per unit is a function of the combined effect of the gamma angle and the workspeed. However, chatter is also a function of the combined effect of the gamma angle and the workspeed. Therefore, since the error resulting from an interrupted surface grows as a function of the error per revolution, there is no advantage in turning the workpiece more rapidly, since it only increases the rate of error generation. By slowing down the workspeed, a higher gamma angle can be used with less risk of chatter, giving a better roundness when the operation has stabilized between error input and roundness correction. (See Fig. 8.19.)

Another problem that is involved in grinding interrupted surface parts is that of eccentricity. During the turning operation of a part, the different diameters turned are reasonably concentric. Normally, if that concentricity is maintained throughout the manufacturing operation of the part, it is adequate for the final function. For example, gear shafts that have a bearing diameter that must be ground.

Unfortunately, many such shafts have keyways machined into the ground diameter, constituting an interrupted surface part. On such parts, since more metal is removed from the side with the interruption than from the side without it, the axis of the diameter ground is shifted relative to the unground diameter and an eccentricity results. That can be quite serious in the case of a gear machined integrally with the shaft whose pitch line must be concentric with the bearing diameter.

Here again, the only way to minimize that generation of eccentricity is to use a very hard regulating wheel, a high gamma angle, and a relatively low workspeed. (A gamma of 8° to 10° and a workspeed of 60 to 70 sfm.)

The best regulating wheel for minimizing out-of-roundness coming

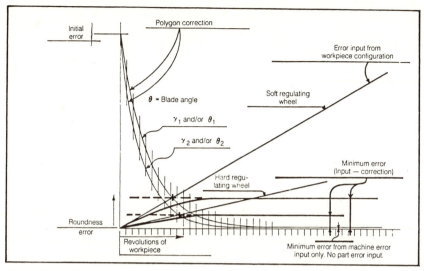

Fig. 8.19   With interrupted surface parts it is advantageous to use a high gamma angle, high blade angle and a low regulating wheel rpm.

from an interrupted surface part is a vitrified wheel. However, vitrified wheels have absolutely no damping, and if the metal removal rate exceeds a rather low value, the wheel will vibrate at a very high frequency that is equal to the resonant frequency of the wheel itself. (The low damping factor of a vitrified wheel is proved each time a wheel is tested for cracks by ringing it with a light wooden hammer.)

If the grinding operation of a part with an interruption is very critical for roundness, it is advisable to rough-grind it with a very hard rubber-bonded regulating wheel, leaving only a minimum of stock for finishing. The finish operation can then be performed with a vitrified wheel, producing the very best roundness possible.

Poor Support.   Poor support of the workpiece during the grind is a very common cause of part out-of-roundness. The fault can be either in the blade or regulating wheel profile. When one is grinding profiled parts on tungsten carbide blades, the blades wear and must be reground. In regrinding, care is not always taken to ensure the proper profile; consequently, the part will rock around a high point during the grinding operation. As the part rocks, 1 end dips toward the blade and the other end lifts off the blade. The end that dips toward the blade is moving down where the wheels are closer together and more metal is removed. The end that is lifting off the blade moves up where the wheels are farther apart and less metal is removed. As the wheel cuts itself free on the low end, the rocking cycle is reversed and the other end dips toward the blade.

Such out-of-roundness is easily identified. Where the part is supported on the blade, it will be round. The unroundness will appear where the part is unsupported, and the lobing will be out of phase half a lobe on the opposite ends.

The same phenomenon occurs if the profile dressed into the regulating wheel does not match the work profile produced by the grinding wheel profile, with the bearing high in the middle. In general, the parts are 5-lobed.

Axial Profile Errors

Axial profile errors constitute 1 of the most common problems encountered on a centerless grinder. In order to solve the problem, one must understand a few basic laws in the grinding process. For example, on a rigid part the dimensions of the different diameters on the workpiece are reproduced directly from the grinding wheel. On a flexible part the dimensions directly across the shoulder are also reproduced from the step trued on the grinding wheel, but between the shoulders the part can bend, usually because of poor support on the regulating wheel. That means that there are basically 5 sources of axial profile errors:

1. The grinding wheel truing cam
2. The grinding wheel truing device
3. The regulating wheel truing cam
4. The regulating wheel truing device
5. The machine slides

The Grinding Wheel Cam

This is the most common source of axial profile errors because the accuracy required in manufacturing a cam is little appreciated. A rule of thumb is that the tolerance produced in the cam must be held to 1/10 of the working tolerance on the workpiece. The reason for that close tolerance is that the cam error is reproduced on the radius of the part, whereas the part tolerance is measured on the diameter. That means that any error of the cam is doubled on the workpiece.

The Grinding Wheel Truing Device

The 2 principal sources of error in the truing device are the longitudinal and transverse slides. The problems of the longitudinal slide usually arise from wear and those of the transverse slide from binding during the displacement, preventing the follower from following the cam.

The Regulating Wheel Truing Cam

Although the regulating wheel truing cam is not required to be as accurate as the grinding wheel truing cam, any errors that exist must not cause the part to bear in the middle. It is always best for the part to bear a little heavier on the ends than in the middle.

The Regulating Wheel Truing Device

Although the regulating wheel is trued much less often than the grinding wheel, its truing device is much more exposed to grit and it, too, is subject to wear. The life of this truing device can be greatly extended by carefully wiping the slides before each truing operation.

The Machine Slides

If the machine slides are poorly fitted (from wear or from being badly adjusted), the rock in the slide can cause straight taper in the part. It cannot affect the axial profile accuracy from step to step.

If an error is perceived in the part profile, it can be determined whether it is in the cam or in the truing device by dressing the wheel in 1 direction and then measuring and recording the error. The wheel is then dressed in the opposite direction, and again the parts are measured and recorded. The parts ground from the dresses in the 2 directions are then compared. If the error in the part remains the same when one is dressing in both directions, it comes from the cam. If the error in the part changes with a change in dressing direction, then it comes from the truing device.

Cylindrical Straightness

On many operations cylindrical straightness is more important than roundness, though a close tolerance for the 1 usually implies a close tolerance for the other.

Infeed Grinding

On infeed grinding, cylindrical straightness is a function of the profile dressed into the wheels, once the best possible taper has been set on the swivel plate. The most common errors are found in profile grinding, because of the difficulty of making grinding and regulating wheel cams. On older machines there is also the problem of truing device wear.

Throughfeed Grinding

On throughfeed operations there can also be some problems with cylindrical straightness. The most common source of problems in holding straightness on throughfeed is in the setting of the guides. They must be set parallel to the axis of the workpiece and in line with the support face of the regulating wheel. The outgoing guide must be set the more accurately of the 2, since the part is finished when it comes out of the rear of the machine onto that guide.

The most accurate method of setting the guides is as follows: First, grind a part with the rough setting of the guides; then cover the straightest half of the length of the part with prussian blue or red lead. Advance the guide until it is well in front of the bearing surface of the regulating wheel. Place the part on the blade and guide with the prepared end hanging over the regulating wheel. Start the regulating wheel and drop the guide back slowly until the contact between the part and the regulating wheel can be felt. Advance the guide until the part has cleared the regulating wheel; then remove the part, taking care not to smear the mark left by the regulating wheel. Examine the regulating-wheel mark to see if the contact of the work on the wheel is a straight line or if it bears heavily at one end or the other of the test length of the sample. Depending on the mark made by the wheel on the workpiece, the guide is swiveled to bring it into a straight line. (See Fig. 8.20.)

In the event that safety regulations will not permit the setup man to hold the part with his hand while the regulating wheel is turning, a spring clip can be made to hold the part in position on the blade and guide while it is brought back against the regulating wheel.

Another source of cylindrical straightness error is what is called the "crowned cam effect." The grinding wheel is trued with a crowned shape when one is throughfeed grinding in order to give an entry into the front of the wheels to remove the metal, a straight portion that represents the sparkout, and a clearance at the back edge of the wheel that keeps the part from cutting on a sharp rear corner. As the part enters at the front of the wheels, the taper on the grinding wheel produces a taper on the front of the workpiece, lifting it off the guide. (See Fig. 8.21.) As the part advances through the wheels, bearing is maintained on the extreme front of the workpiece and at the initial point of contact between the grinding wheel and the workpiece, making a hollow profile between the 2 contact points; thus the part is hollow in the middle.

The error from the crowned cam effect is a function of the stock removal per unit of width of the grinding wheel; therefore, it can be eliminated, or at least considerably reduced, by progressively reducing the stock removal per pass or by grinding the part on a machine mounting wider wheels.

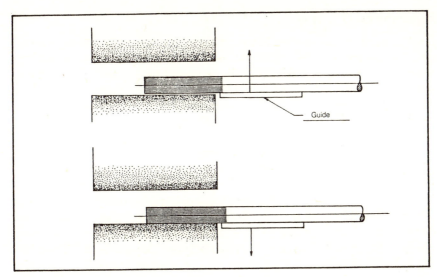

Fig. 8.20   Aligning guides.

Fig. 8.21   "Crowned cam" phenomenon generates parts low in the middle.

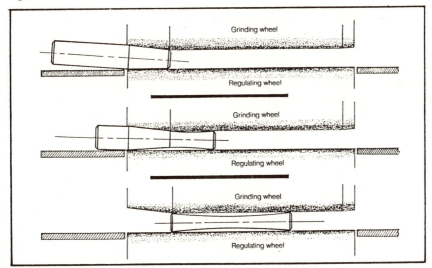

Concentricity

Although roundness is by far the most important element of workpiece quality that must be considered on a centerless grinder, still there are occasional concentricity requirements that necessitate a thorough understanding of the concentricity ramifications of the process. First of all, we must redefine the concept of concentricity here in order to appreciate how it is affected by the various elements of a centerless grinding setup.

Just as true roundness can be defined as being a circle, all points of which are equidistant from a single point, the center, concentricity can be defined as 2 circles, all points of which on the 1 circle are equidistant from a single point, the center, and all points of which on the other circle are equidistant from the same center point but at a different radial distance. (See Fig. 8.22.)

Although that is an accurate definition geometrically, it cannot be used for measuring concentricity on a centerless ground part, for the simple reason that there is no physically defined center of the part that can be used as a reference for both circles. Consequently, to determine whether or not 2 diameters on a workpiece are concentric, it is necessary to use 1 diameter as the reference for establishing the axis and then measure the radial difference around the circumference between the 2 diameters.

That means that a more accurate definition of concentricity for a centerless ground part is "a part with 2 diameters having a constant radial relationship around the circumference and along the common axis." (See Fig. 8.23.)

There is a law of grinding that states that all diameters on a single shaft that are ground simultaneously must be concentric according to the second definition given above. In that definition no reference is made to a common center of the 2 diameters for the simple reason that the common center is the basis for the definition for roundness, and it is very important in considering centerless grinding problems to keep separate the problem of concentricity from that of roundness. Consequently, by that definition, a part can be out-of-round, tapered, or chattered, but if the diameters have been ground simultaneously, *they will be concentric*! There are 2 reasons for that, 1 of which is that there can be only 1 axis of rotation of a part at any given moment, and the other is that on parts with stepped profiles, the radial difference of the wheel profile must be constant around the wheel, and since it is reproduced into the workpiece, the radial difference on the step profile around the workpiece circumference must also be constant.

There is, however, no assurance that a diameter that is not ground simultaneously with the others will be concentric with them. Whatever

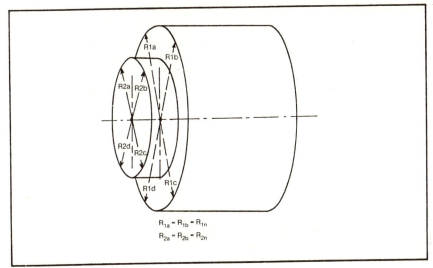

Fig. 8.22   Concentricity, a definition for centertypes.

Fig. 8.23   Concentricity, a definition for centerless.

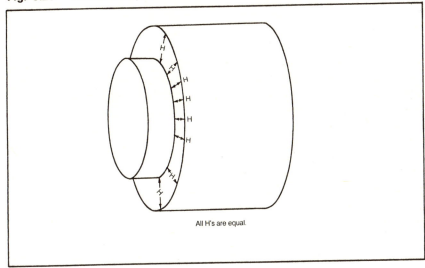

initial concentricity exists will remain reasonably constant, unless there are extraneous influences (usually in the part configuration itself) that will affect that concentricity.

Nevertheless, it is possible to improve the concentricity between a diameter being ground and a diameter that is not being ground, on condition that a way is found to introduce the unground diameter into the setup as part of the process. One can do that either by supporting a previously ground operation on the blade and regulating wheel while grinding the other diameter, or diameters, (See Fig. 8.24.) in which case the tolerance of the diameter bearing on the regulating wheel must be added to the working tolerance of the diameter being ground—or the diameter being ground must be supported on the regulating wheel while the previously finished diameters are supported on the blade. (See Fig. 8.25.)

Bow Correction

Anyone acquainted with the centerless grinding process is quite aware of the fact that a bowed part can be ground straight if the length-to-diameter ratio is such that the resistance to bending of the part is greater than the cut force applied by the grinding wheel. What is less well known is that it is also possible to correct bow on a part whose length-to-diameter ratio is such that its resistance to bending is less than the cut pressure of the grinding wheel. In that case the workpiece is pressed against the regulating wheel over its entire length during the grind, which would seem to indicate that the same amount of metal would be removed from all portions of the part surface, with the result that no correction would occur. There is, however, another variable in that process which is extremely important in bow correction—the resilience of the regulating wheel.

Over the years it has been learned that vitrified wheels are inapplicable to the function of a regulating wheel used for normal grinding operations because of the inherent lack of damping, which results in very high sensitivity to resonant chatter between regulating wheel and workpiece, particularly when one is using a vitrified grinding wheel. Therefore, rubber-bonded regulating wheels are almost universally used for that purpose because of their high damping characteristics. However, there is an inverse relationship between damping and hardness, and the wheel that is chosen for best damping is sufficiently resilient to permit a significant penetration of the workpiece into the surface of the regulating wheel during a normal grinding operation, and the depth of penetration is roughly proportional to the radial grinding force.

Fig. 8.24   Concentricity by
reproduction

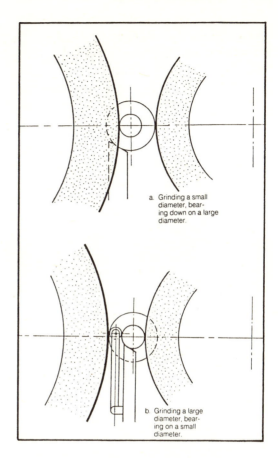

a. Grinding a small
diameter, bear-
ing down on a large
diameter.

b. Grinding a large
diameter, bear-
ing on a small
diameter.

Fig. 8.25   Concentricity generation without compromising size.

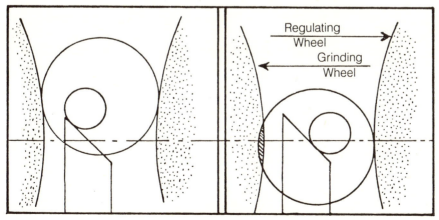

In the case of a part that, in the free state, is bowed but tends to be straightened under the grinding pressure pushing it against the regulating wheel, the force that the workpiece applies to the regulating wheel at each point along its axis is a function of the radial grinding pressure plus the internal stresses created in the constrained part by its tendency to reassume its free-state form. In the case of a workpiece having a long uniform bow from 1 end to the other (See Fig. 8.26.) when the high side of the bow is to the grinding wheel side, the workpiece applies a pressure to the regulating wheel diametrically opposite that is equal to the radial cut force *minus* the bending stress. At the ends of the part, the force applied to the regulating wheel is equal to the radial cut pressure *plus* the bending stress.

What that means is that wherever the support pressure is greatest between the work and the regulating wheel, the axis is shifted toward the regulating wheel and less metal is removed from the part at that time. Wherever the force is least between the workpiece and the regulating wheel, there is less deflection of the regulating-wheel surface and the workpiece is held in closer contact with the grinding wheel, removing more metal. Consequently, the changing amount of penetration of the work surface into the regulating wheel, synchronized with the presentation of the high side of the bow to the grinding wheel, produces an effect of grinding more off the high side than off the low side, providing

Fig. 8.26   Bow can correct without the part leaving the regulating wheel.

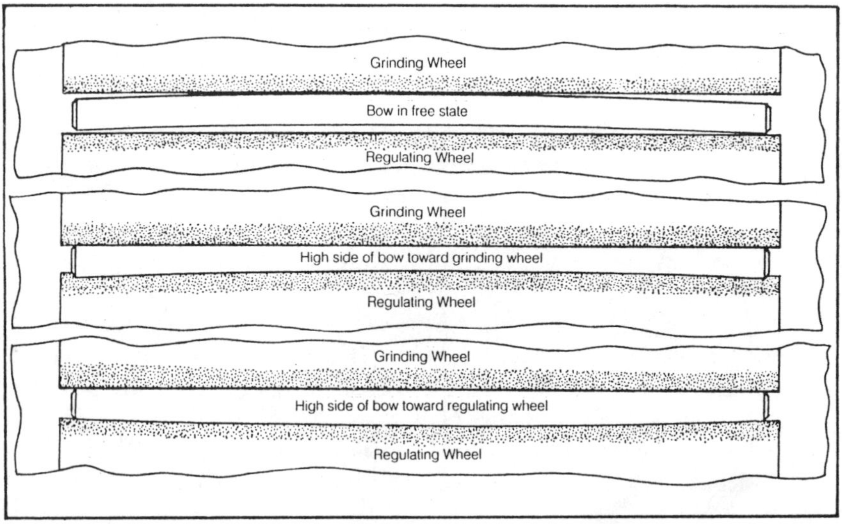

a bow correction effect even though the part has never left contact with the regulating wheel. A simple conclusion to be drawn from this analysis is that for maximum bow correction, it is necessary to use the softest possible regulating wheel.

When grinding a bowed part on a center type grinder, it is necessary to remove stock equal to the total indicator reading of the runout in order to clean it up. On a centerless grinder, it is necessary to remove stock equal to only half the total indicator reading in order to clean it up.

Grinding Thin-Walled Parts

Another area of centerless grinding applications that is fairly common and has very special problems is that of grinding thin-walled parts. That category of part configuration encompasses 2 well-known parts, bearing races and cylinder liners. In both of those categories the words *thin wall* are applicable to only a portion of the field in each category. On those parts where the difference between the inside diameter and the outside diameter is such that there is sufficient rigidity to the part for it to act as if it were a solid part, it no longer qualifies as a thin-wall part. Actually, there are relatively few bearing races that fit that category, but when they do, they are extremely difficult to grind. In the field of cylinder liner grinding, there are 2 types of liners in current use: wet liners and dry liners. The wet liner is ground on only a few rather narrow bands along the diameter, and the wall thickness is such that their behavior under grind is almost identical to that of a solid part. Dry liners, however, are ground over their entire surface, and since functionally the thinner the wall of a dry liner the more efficiently it transmits heat from the explosion chamber to the cooling medium, the more the trend is to decrease their wall thickness. In addition to that, current developments in diesel engine design are causing wet liner design to be replaced by dry liner design because of its more efficient cooling. Since more dry liners will be made, and they will probably become thinner and thinner walled, it is very important for us to understand all of the problems involved in their grinding process.

The big problem with grinding these parts is the extremely low resistance to deflection resulting from the radial cut pressure of the grinding wheel. Considerable work has been done in mechanically reinforcing the parts by applying an overhead pressure roller, with quite good results as far as permitting the use of a relatively high metal removal rate is concerned. (Fig. 8.27.)

Fig. 8.27   Setup for grinding thin wall parts.

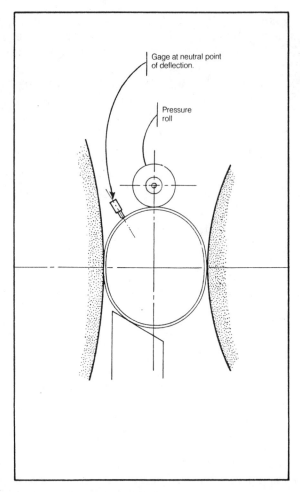

Gage at neutral point of deflection.

Pressure roll

However, any additional contracts added to the part act as additional elements to the "framing figure" and tend to change the roundness generation. Therefore, there is a limit to the amount of force that can be applied on those additional contacts and still be able to generate a round part.

CHAPTER **9**

Vertical-Spindle Surface Grinding

D. H. Youden

Process Characteristics

Vertical-spindle surface grinding is often thought of as exclusively a roughing operation. While the process is a common first operation on a casting, forging, or weldment, it is also used for fine finishing applications that demand close control of flatness and surface finish. Because the abrasive wheel[3] on those machines are nearly always wider than the work, they are not mounted on cross slides or other devices for the purpose of traversing the wheel across the work. The elimination of those mechanisms contributes to the accuracy of the machine tool.

Grinders of that type usually are equipped with wheel drive motors, which seem to have very high horsepower ratings when compared with other types of grinders. That is because the machines are designed to cut using the entire end surface of the abrasive. The large area of that abrasive face requires a high power wheel drive, even though the cutting forces per unit of area are no higher than in other grinding processes. Fig. 9.1 illustrates a large high powered machine.

Flatness

One unique characteristic of vertical-spindle surface grinding is that the geometry of the work, i.e., its flatness, is generated by the wheel surface and does not rely on a shape that is formed into the wheel by a diamond or other dressing device. That means that, assuming the wheel is properly chosen for the work being done, flatness is not lost as the wheel

233

Fig. 9.1   A large machine that can rotate workpieces 10 ft. in diameter. The wheel on this machine is driven by a 250-horsepower motor.

wears. If the wheel dulls abnormally, forces can be generated that will alter the machine geometry and flatness will be lost, just as on other types of surface grinders.

Surface Finish

Vertical-spindle surface grinding leaves the workpiece with a characteristic cross-hatched pattern on the finished surface. The depth of that pattern is a function of wheel selection, feedrate, sparkout time, and work material. Finishes as good as 1 or 2 microinches RMS can be obtained if proper attention is paid to all of those variables.

Individual abrasive grains leave very long scratches on surfaces produced by vertical spindle surface grinding. That may be viewed as a process flaw or as an advantage depending on the end use or purpose of the ground surface. Fig. 9.2 illustrates the characteristic pattern resulting from vertical-spindle surface grinding.

Fig. 9.2 The characteristic surface pattern produced by rotary vertical-spindle surface grinding can be seen on these large steel billets.

Self-Dressing of the Abrasive

The abrasive wheels used for this type of grinding are designed to wear at a relatively rapid rate. It is that wearing away which continually exposes new, sharp abrasive grains on the working surface of the wheel. If the grinding wheel is too hard for the application, it will not wear away under the forces generated by the grinding action. Under those conditions the wheel has to be broken down periodically by use of a dressing device, which often consists of a group of star-shaped steel wheels, which are free to rotate on a pin. That mechanism is moved across the working face of the rotating wheel, causing small areas of high stress that break the bonds between the abrasive grains, thus resharpening the wheel.

If the wheel is too soft for the job at hand, it will break down prematurely and the force between the wheel and the work will not rise to a level sufficient to cause the wheel to cut the work. The symptoms of that condition are high wheelwear and good surface finish coupled with low readings on the machine load meter. The relationship between power, G ratio and feed rate are shown in Fig. 9.3.

Use of the Load Meter

Most vertical-spindle surface grinders have a load meter, a device that indicates how much power is being consumed by the motor, which ro-

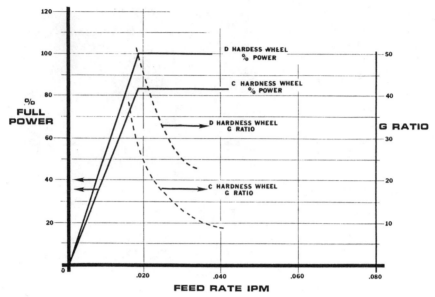

Fig. 9.3   How machine power and G ratio vary with wheel hardness and feedrate.

tates the grinding wheel. Some machines use a simple ammeter for that purpose while others have a true wattmeter. The advantage of the watt-meter is that it is more sensitive to light loads and to small changes in load than the ammeter, and so it is more often found on small machines. The disadvantages of the wattmeter are complexity and cost.

The load meter is most useful in determining how the grinding wheel is performing. A wheel that is too soft for the job will not allow the machine load to build to an adequate level no matter what rate of feed is selected. An excessively hard wheel, on the other hand, will cause the meter to indicate a very high load, which can be reduced only by dress-ing the grinding wheel. If the wheel hardness is properly matched to the job, the load meter will indicate nearly a 100 percent load and its reading will vary slightly over the course of a minute or so as the wheel dulls and then dresses itself.

If the infeed of the grinder is shut off and the wheel is cutting freely, the load meter will drop to its idle reading quickly, indicating that the wheel is sharp and has sparked out. A hard or dull wheel will cause the load meter reading to drop slowly or not at all.

Stock Removal

Vertical-spindle surface grinding is most often used as a first step in producing a machined part. When the process is used in that manner, the amount of material removed can be very large. A big machine can easily remove 1,000 cubic inches of material per hour from an iron casting. The same machine, properly maintained, can produce a surface flat within .0005 in. per foot.

The amount of stock allowed for grinding will, of course, vary with the size of the workpiece, its surface condition, and the purpose of the grinding operation. A large fabrication or casting may have a grinding allowance of as much as ½ in. while a more delicate part may require the removal of only .002.

Types of Machines

Reciprocating Table

Reciprocating vertical spindle grinders are made in 2 basic types, those upon which the workpiece is moved back and forth beneath the grinding wheel and those where the wheelhead and its downfeed mechanism are reciprocated above the stationary workpiece. The more common design is that which moves the work as shown in Fig. 9.4. Reciprocating grinders can accommodate long workpieces more easily than rotary table machines can, and they offer simplicity of construction and potentially greater accuracy than do machines that grind with the periphery of the wheel.

Moving head grinders of that type are commonly employed for the grinding of large knife blades, which are fixed on a tilting electromagnetic chuck mounted low between the ways that guide the wheelhead during reciprocation. Those machines can handle very large workpieces without sacrificing accuracy and ridgidity.

Rotary Table

The rotary table vertical-spindle surface grinder is perhaps the most common configuration of vertical-spindle surface grinder. It is characterized by a flat, circular worktable, that is usually mounted on a linear slide so that the work may be moved out from under the grinding wheel for inspection and handling. The worktable is sized so that its entire surface may be ground flat using the machine's own grinding wheel.

Rotary table machines are manufactured in a very wide range of sizes. Worktables vary from 12 in. in diameter to at least 140 in. Wheels, which may be cylinders or of segmented construction, range from six in. to more than six ft. in diameter. The largest of those machines can accept work up to 6 ft. high.

All rotary table grinders of that type have some sort of mechanism for

Fig. 9.4   A typical, modern vertical-spindle surface grinder.

adjusting the parallelism of the wheel spindle and the worktable centerline. It is only when those 2 axes of rotation are parallel that the machine will produce a perfectly flat surface.

The surfaces produced on rotary table surface grinders have a characteristic pattern of intersecting arcuate scratches, which appear to go
outward from the point on the work located at the center of the worktable. The relative depth of the scratches made by the forward edge of
the wheel and those made by the trailing edge can be used as a general
indication of the flatness of the ground surface. If those scratches are not
of equal depth, it can be assumed that the axis of the wheel was tilted in
relation to the work.

Single-Pass, Multiple Head

Single-pass or throughfeed vertical-spindle surface grinders are used where production rates are very high and a machine can be dedicated to producing a single part or a family of similar parts as illustrated in Fig. 9.5. Automotive connecting rods and main crankshaft bearing caps are examples of parts that are processed on throughfeed surface grinders. Those machines may have a single-wheel spindle, but they are more commonly built with 2 to 5 separate spindles, each mounted on its own slide. The action of those individual slides is controlled by a series of fingers that sense the machined surface of the workpiece just after it passes under the wheel, which is mounted on the slide to be controlled. If the wheel has worn by some preset amount, the part surface will be enough higher than usual to deflect the finger. (see Fig. 9.6) This causes a switch to signal the slide to move toward the workpiece by an amount

Fig. 9.5 A single-spindle, throughfeed grinder tooled for grinding small bearing races.

Fig. 9.6   A group of work pieces mounted on the magnetic chuck of an average grinder. Note the in-process, work riding gage positioned to control the finished size of the work.

calculated to compensate for the detected wheelwear. That process operates independently on each of the wheel slides, causing them to periodically advance until 1 of the grinding wheels is worn to the point where the machine must be stopped for service.

The workpiece on throughfeed vertical-spindle surface grinder does not usually rotate but instead is clamped into a fixture, which carries it around the periphery of the machine, passing in sequence under each of the grinding wheels. Each wheel is set to remove a portion of the total stock to be removed so that the part is brought to finished size in a series of steps.

Work Holding Methods

Electromagnetic Chucks

The most common method of holding the workpiece on a vertical-spindle surface grinder is by use of an electromagnetic chuck, as shown in Fig. 9.7. Many machines include that type of chuck as an integral part

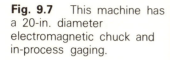

Fig. 9.7 This machine has a 20-in. diameter electromagnetic chuck and in-process gaging.

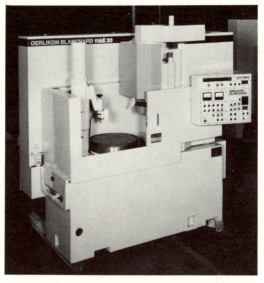

of the worktable. Those chucks are available in several different configurations, each suited to certain types of workpieces. However, the basic principal of their operation is always the same. An electric current is passed through a series of coils of wire, creating a magnetic field. The internal machining in the chuck causes that field to pass through the surface of the chuck and, in turn, through the workpiece or pieces that are placed on it. It is important that the surface of the chuck be kept smooth, flat, and clean to ensure that the maximum holding force is available at all times.

When the machining operation is finished, electromagnetic chucks must be demagnetized. That is done by reversing the polarity of the magnetic field a number of times. Each time the polarity is reversed the intensity of the field is reduced. After a dozen or so reversals the work may be removed from the chuck.

It is possible to use magnetic chucks to hold parts that are not made of iron or steel. To hold, those parts must be blocked in place, with pieces of magnetic material being spaced around them so that they cannot move during the grinding operation. The blocking material must not protrude above the surface to be ground, or it will interfere with the operation.

Permanent Magnet Chucks

Permanent magnet type chucks are more commonly found on smaller sizes of vertical-spindle grinders. Recent improvements in magnet materials have made that type of chuck a good alternative to the more expensive electric chucks. The holding power of permanent magnet chucks is equivalent to their electric counterparts, but that power is fixed, a disadvantage that can be of significant importance in operations where good flatness must be achieved on relatively thin, flexible parts.

Vacuum Chucks

Vacuum chucks may be fabricated with porous ceramics or metallic materials, or they may be solid with a grooved top surface and distribution holes for removing the air from between the chuck and the workpiece.

Porous.   Porous vacuum chucks can be made so that the porous area exactly fits the workpiece or in a larger, more general shape. If the chuck's working area extends beyond the edges of the work, its surface must be masked to prevent leakage of air and coolant into the surface of the chuck. If air leaks into the chuck, it will greatly reduce the holding power. Coolant, which carries grinding debris, will cause clogging of the minute passages in the porous material. Some porous chucks must be periodically back flushed with air or water to wash out bits of dirt that lodge in those passages.

Grooved or ported.   Grooved or ported vacuum chucks are less sensitive to dirt than porous chucks but do not offer uniform support to the work. When one is grinding very thin parts, the pattern of grooves will sometimes "print through" and be apparent on the workpiece as errors in flatness. Those chucks are made by machining a series of grooves or holes into the flat working face of the chuck and connecting them to a common vacuum source. Usually each groove or port can be individually disconnected by inserting a small plug into the face of the chuck. That feature minimizes the amount of masking required.

Fixtures

When the part to be ground does not have a bottom surface that can support it in a stable position, a fixture is required. For vertical-spindle surface grinding the fixture is usually very simple. If the part is one of a kind or is processed in limited quantities, the fixturing is commonly made up of a series of blocks or plates of scrap steel placed under and around the work to support it in the desired attitude and to keep it from sliding across the surface of the chuck.

If the workpiece is processed in large quantities, the fixture may be somewhat more elaborate. It may include hardened locating surfaces and manually or power-operated clamping mechanisms. Often those fixtures are held to the grinder with the machine's magnetic chuck used as the clamping device.

Abrasive Selection

Effect of Workpiece Material

As in all grinding processes, the work material affects the selection of abrasive. Ferrous materials are cut most effectively with mixtures of aluminum oxide grains held in vitrified bond. Soft, nonferrous metals, such as aluminum and the copper metals, are best cut with silicon carbide abrasive or a blend of silicon carbide and aluminum oxide. Glass can be ground with silicon carbide, but quartz, tungsten carbide, sapphire, and other very hard materials require diamond abrasive wheels. Diamond wheels may utilize a resin-bonded construction, or the diamond particles may be held in a metallic matrix. The choice of wheel construction is very much dependent on the characteristics of the individual job and should be discussed with the supplier of the abrasive.

Self-Dressing

Vertical-spindle surface grinders are most effective when they are used in a manner that causes the grinding wheel to sharpen itself by breaking down in a controllable fashion. As the feedrate of the machine is increased, the load meter will indicate an increasing power level up to the point where the wheel begins to break down. Increasing the feedrate beyond that point will not cause the power to increase further but will only cause more rapid breakdown of the abrasive wheel. The feedrate at which breakdown occurs can be changed by changing the traverse or rotational speed of the work. Power consumption at the breakdown point can most effectively be controlled by wheel selection. However, some grinders tilt the wheel slightly during rough grinding to present a smaller working face to the work, thus increasing the local power density and causing wheel breakdown. A machine operating in the self-sharpening mode is characterized by a slow pulsation of the load meter as the wheel dulls and then breaks down, exposing new, sharp abrasive to the cutting process.

The grinding ratio, or G ratio (see chapter 2), is a common term that has perhaps more significance in this type of grinding than in some others because of the use of soft wheels, which wear rapidly. The G ratio is the volume of workpiece material removed divided by the volume of grinding wheel worn away while removing that material. At low feedrates the G ratio is usually a large number. As the feedrate is increased,

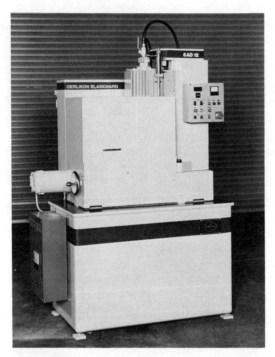

Fig. 9.8   A very small machine that has the capability to use cubic boron nitride abrasive effectively. The grinding wheel on this machine is 8 in. in diameter.

the G ratio decreases gradually at first. When the forces become large enough and the wheel begins to self-sharpen, then it decreases very rapidly. Typically, the G ratio on a properly operated vertical-spindle surface grinder will be between 5 and 20 when one is using aluminum oxide abrasive to grind ferrous material.

Use of Cubic Boron Nitride

In recent years cubic boron nitride has seen increased use as an abrasive. That very hard material works best at speeds beyond the reach of most vertical-spindle surface grinders. Cubic boron nitride has, however, been successfully used on at least 1 small vertical grinder (Fig. 9.8) that has the capability of reaching a wheel surface speed of nearly 8,000 ft. per minute. At speeds below 6,000 ft. per minute much of the advantages of the unusual abrasive are lost.

Machine Operation

Ferrous Metals

Iron and steel are usually ground with an electromagnetic chuck. Before one places the work on the chuck, the surface of the chuck must be cleaned and dried of any coolant. After cleaning the chuck, it is good practice to lightly stone its surface with a soft abrasive to remove any burrs or raised areas that have resulted from contact with the previous load of parts.

Once the chuck is clean, the new load of workpieces can be positioned on it. Parts should be placed around the periphery of a rotary chuck and should be in firm contact with one another or with steel blocking plates. Special care must be taken to be sure that all parts are in good contact with the chuck surface. Shims should be used so that no part can rock or wobble during grinding.

When the chuck is loaded with parts, move it beneath the grinding wheel and, without starting the wheel, position the wheel so that it just clears the highest part on the chuck. Close the machine guards, start the chuck in motion, and then start the wheel rotation. Bring the wheel into contact with the work carefully, watching for the initial sparks while the wheel is lowered. Once contact has been made, turn on the coolant and start the power feed. The feedrate and distance should be set appropriately for the job at hand.

Nonmetallics

The most notable common denominator in the grinding of nonmetallics is the work holding fixturing. Vacuum chucks are perhaps the most prevalent devices employed; however, many parts are held by being secured to steel or iron plates, which are in turn held on magnetic chucks. Wax and pitch are the most common adhesives used to hold those workpieces even though they both require heat to be used in the process.

Semiconductor Materials.   Integrated circuit and transistor substrates of silicon, germanium, or sapphire are often ground to their finished thickness with vertical-spindle surface grinders. The machines used for that operation usually are rotary table types, although there are some throughfeed machines available suitable for that application. Nearly all grind on more than 1 workpiece at a time, either by using multiple part holding fixtures on a large rotary table or by employing several wheel spindles, each machining a single part mounted concentrically on a small, independent work table.

It is especially important when one is grinding these very brittle materials to reduce any vibration in the equipment to an absolute minimum level. Wheels and wheel spindles must be balanced to precision tolerances, and the balance must be maintained throughout the life of the wheel. Vibration from other sources, such as the table rotational drive, coolant pumps, and hydraulic systems, must be minimized.

Cleanliness is also important, as microscopic bits of abrasive, work material, and dust between the workpiece and the fixture can lead not only to out-of-tolerance parts but also to actual mechanical failure of the workpiece material. That failure manifests itself as small star-shaped cracks, edge chipping, or, in severe cases, as breakage of the workpiece.

The grinding wheels used for grinding semiconductors are normally resin-bonded diamond wheels with grit sizes between 200 and 1,200. Concentration values are generally near 50. The working face of the wheel must be very accurately trued after the wheel is mounted in the machine. One method of achieving the required truth is to lap the wheel against a glass plate using diamond lapping paste. The glass plate is mounted on the worktable and rotated at a slow rate. With the table running, the wheel spindle is lowered with the wheel motor turned off until the wheel is in contact with the abrasive charged plate. The wheel will rotate slowly, driven by the friction between it and the plate. The machine is allowed to run in this way for 10 minutes or more until the desired wheel surface condition is obtained.

Coolants employed for that operation are water-based synthetics although deionized water without any additives is also commonly used. The use of deionized water demands that special precautions be taken to prevent corrosion of the machine and the fixture.

Economics of Vertical-Spindle Surface Grinding

A vertical-spindle surface grinder, like any other machine tool, consumes several commodities in order to remove metal. If we total the cost of those commodities and divide that total by the number of cubic inches of metal removed, we can generate the cost of removing 1 cu. in. of metal.

The commodities consumed by a grinding machine include time, abrasive, coolant, and electric power. The cost of coolant is generally small compared with the other factors, and it will be neglected in the following analysis. With the cost of coolant neglected, the cost of metal removal can be stated as:

$$C = (Ca+T\times Ct+Ce\times Km\times Vm)/Vm \tag{9.1}$$

Where

Ca = cost of abrasive, $
Ct = Cost of time, $ /Min
Ce = Cost of electricity, $ /KWH

Vm = Volume of metal removed, in.^{3}
C = Total cost, \$ /in.^{3}
T = Time in cut, min
Km = Material constant, KWH/ in.^{3}

Km is dependent on the material being ground. Its value is about 0.167 for steel and 0.067 for cast iron.

The cost of abrasive, Ca, is composed of 2 parts, the purchase price of the material and the cost of handling it. In order to compute the cost of abrasive used during a single grinding test, we can divide the total cost of 1 wheel or set of segments by the usable volume of abrasive contained therein and then multiply the result by the number of cubic inches worn away during the grinding test.

$$Ca = Cs/Vs \times Va \qquad (9.2)$$

Where:

$$Cs = Cm + Ch \qquad (9.3)$$

And:

Cs = Cost of one set of segments, \$
Cm = Cost of abrasive material, \$
Ch = Cost of handling, \$
Vs = Usable volume of abrasive, in.^{3}
Va = Volume of abrasive consumed, in.^{3}

The cost of 1 set of segments is simply the purchase price of 1 segment multiplied by the number of segments in a set. The cost of handling the segments is composed of the cost of the time to install them and, if the segments are blocked down in the chuck to obtain additional life, the cost of installing the blocks. The cost of installing the blocks is nearly equal to the cost of installing the segments in the first place, as the operations are nearly identical. We may then rewrite the cost of the segments as:

$$CS = [Ps \times Ns + (Nb+1) \times Ts \times Ct] \qquad (9.4)$$

Where:

Ps = Price of one segment, \$
Ns = Number of segments
Nb = Number os sets of blocks used
Ts = Time to change segments, min

The usable volume of abrasive, Vs, may be calculated by multiplying the area of 1 segment, As, by the usable length of the segment, Lu, and the number or segments.

$$Vs = As \times Lu \times Ns \qquad (9.5)$$

The area of the segment is obtainable from the abrasive specification sheet supplied by the abrasive manufacturer.

The final expression for the cost of abrasive may then be written:

$$Ca = [Ps \times Ns + (Nb+1) \times Ts \times Ct] / (As \times Lu \times Ns) \times Va \qquad (9.6)$$

The cost of time, Ct, is expressed in dollars per minute and is made up of 2 parts, the cost of the machine tool and the cost of the operator. For our purposes it will be sufficient to use the simplest expression possible to compute the cost of the machine in dollars per minute. We will amortize the purchase price over the life of the machine:

$$Cmt = Pmt / (60 \times Lmt) \qquad (9.7)$$

Where:

Pmt = Purchase price of the machine,
dollars
Lmt = Life of the machine, hours

There are many, more complicated, methods for expressing that quantity, and any of them may be substituted for that expression.

The cost of labor is simply the hourly labor rate, Cl, charged to the operation divided by 60 to obtain a rate in dollars per minute. Ct can therefore be expressed as:

$$Ct = Pmt / (60 \times Lmt) + CI / 60 \qquad (9.8)$$

The time in the cut, T, can be calculated by dividing the amount of downfeed, D, by the feed rate, R:

$$T = D/R \qquad (9.9)$$

Now that all of the variables are defined, we may rewrite the original equation:

$$C = [[Ps \times Ns + (Nb+1) \times Ts \times Ct \times Va] /(As \times Lu \times Ns) + D \times Ct\ R + Ce \times Km \times Vmr]\ 4Vm \qquad (9.10)$$

or:

$$C = [Ps \times Ns + (Nb+1) \times Ts \times Ct \times Va] /(As \times Lu \times Ns \times Vm) + D \times Ct\ /(R \times Vm) + Ce \times Km \qquad (9.11)$$

Substituting the G ratio, G, for Vm/Va, we obtain the final expression for the cost of metal removal:

$$C = [\,Ps \times Ns + (Nb+1) \times Ts \times Ct\,] / (As \times Lu \times Ns \times G) + D \times Ct \,/)R \times Vm) + Ce \times Km$$

$$(9.12)$$

For any specific machine tool many of those values become constants. For a typical medium-sized rotary table surface grinder the values might be:

Ps = $12.50	Lu = 4 in.
Ns = 5	Pmt = $150,000
Nb = 1	Lmt = 40,000 hrs
Ts = 20 min	Cl = $25.00 / hr
As = 18.4 in²	Ce = $0.08 /kwh

using those values:

$$Cg = \$0.479 \,/ \text{min}$$

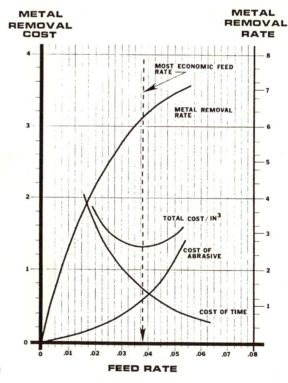

Fig. 9.9   The effect of feedrate on the cost of removing metal with a vertical-spindle surface grinder.

METAL REMOVAL RATE AND COST

AS A FUNCTION OF FEED RATE

And, if the work material is steel:

$$C = .222 / G + D \times .479 / (R \times Vm) + .013 \qquad (9.13)$$

It should be noted that as the rate of downfeed, R, is increased the value of G generally decreases. That causes the value of the first term of the expression to increase with increasing feedrate while the value of the second term decreases. That characteristic is important because the expression has a minimum value at some feedrate. There is, therefore, a feedrate that will minimize the cost of removing metal. If the feedrate is increased or decreased from that optimum value, the cost of removing metal will rise. Fig. 9.9 illustrates how the cost of metal removal and the rate of metal removal vary as the rate of downfeed is changed.

We should take special note of the fact that when a vertical spindle surface grinder is operated with less than a full chuck load, the cost of removing metal will go up roughly in proportion to the percentage of time that the machine is feeding without contact between the wheel and the work. During that time the operator and the machine are consuming time without removing metal.

Also, for the purpose of this analysis, the machine has been assumed to be operating at 100% efficiency except for abrasive changes. In reality an operating efficiency of about 70% would seem reasonable. That decrease in efficiency has the effect of increasing the cost of time, Ct.

CHAPTER 10

Reciprocating Surface Grinding

Robert S. Hahn
Hahn Associates

Introduction

Reciprocating surface grinders are used to move the wheel with rectilinear motion relative to the workpiece to generate a plane surface or a profiled surface with zero curvature in the direction of motion. The wheel strokes back and forth over the workpiece, making "conventional" or up-cutting strokes in one direction and "climb"-cutting strokes in the return direction. The principles of reciprocating surface grinding are closely related to the principles of cylindrical grinding outlined in Chapters 1 and 2. It is suggested that those chapters be read before proceeding with this chapter. There are, however, some unique features of reciprocating surface grinding that are developed below.

Principles of Reciprocating Surface Grinding

The general metal-removal relation for cylindrical grinding operations is presented in Eq. 1.11. Using that relation to obtain Eq. (1.13):

$$\bar{v}_w = \frac{WRP}{\pi D_w} (F_n' - F_{th}')^*$$

(**10.1**)

which gives the plunge-grinding velocity $\bar{v}_w$ as a function of the interface normal force per unit width of contact F'_n. Dividing both sides of that equation by the work rotational speed N_w yields:

$$\frac{\bar{v}_w}{N_w} = \frac{WRP}{\pi D_w N_w} (F'_n - F'_{th}) \tag{10.2}$$

The left-hand term is recognized as the "wheel depth-of-cut" h.

The denominator of the right-hand side is recognized as the work surface speed V_w. Therefore:

$$h = \frac{WRP}{V_w} (F'_n - F'_{th}) \tag{10.3}$$

That equation applies to surface grinding and gives the wheel depth-of-cut h in terms of normal force intensity F'_n. The "work removal parameter," WRP, applies to internal, external and surface grinding with the understanding that it varies, somewhat, with the "Equivalent Diameter" D_e. See (Eq. 1.6). For surface grinding:

$$D_e = D_s \tag{10.4}$$

where D_s is the diameter of the grinding wheel.

The work removal parameter is described in Chapter 1 and also in Chapter 2. If one uses Eqs. (10.3) and (10.4), many of the relations presented in Chapters 1 and 2 for cylindrical grinding apply also to surface grinding.

Most surface-grinding machines are provided with an incremental feed that takes place at the end of each stroke (called the downfeed d). As the wheel engages the workpiece, normal force is developed, causing the system to deflect. Accordingly, the wheel depth-of-cut h is less than the downfeed d, as illustrated in Fig. 10.1. K_m represents the overall system rigidity. (See chapter 7 for a discussion of system rigidity.) The elastic force F_e generated in the system is:

$$F_e = K_m (d - h) \tag{10.5}$$

The normal component of the grinding force F_n required to remove the depth-of-cut h is:

$$F_n = K_c h \tag{10.6}$$

where K_c is the "cutting stiffness." (See Eq. (1.19).) Since those 2 forces are equal and opposed:

$$K_c h = K_m (d - h) \tag{10.7}$$

---

* See Nomenclature in Chapter 1.

Fig. 10.1 Illustration of the wheel depth-of-cut h, the downfeed d, and the system deflection (d−h).

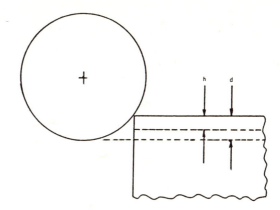

Solving for h gives:

$$h = \frac{K_m}{K_m + K_c} d \tag{10.8}$$

When $K_c \ll K_m$:

$$h \cong d \tag{10.9}$$

However, K_c in practical grinding machines is usually large compared with K_m: thus, when:

$$K_c \gg K_m$$

$$h = \frac{K_m}{K_c} d \tag{10.10}$$

From Eq. 19, Chapter 1:

$$K_c = \frac{V_w W}{WRP} \tag{10.11}$$

hence:

$$h = \frac{K_m d\ WRP}{V_w W} \tag{10.12}$$

That equation shows that the wheel depth-of-cut h is inversely proportional to the work surface speed V_w, and, as a result, the volumetric stock-removal rate Z_w:

$$Z_w = hwV_w = K_m d\ WRP \tag{10.13}$$

is essentially independent of workspeed. That equation is valid for

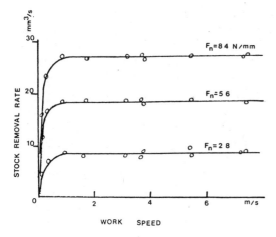

Fig. 10.2   Stock-removal rate vs. work speed (2) AISI 4150, 47-52 R_c, internal grind; D_w = 90mm, D_s = 70mm; Wheel 60L4; Wheelspeed = 30m/sec

workspeeds above a certain value, as demonstrated by the horizontal portion of the curves shown in Fig. 10.2. The steep sections of those curves correspond to the "creep-feed" grinding region where the stock-removal rate is directly proportional to workspeed. That is discussed further below.

The induced force, F_n, can be found by combining Eqs. 10.6 and 10.8 thus:

$$F_n = \frac{K_c K_m}{K_c + K_m} d \qquad (10.14)$$

Since surface finish and wheelwear are dependent upon F_n, that equation shows that the induced force for a given downfeed, d, will vary as the system rigidity varies. It explains why a wheel of given specification may behave differently on machines of differing rigidity.

Equations (10.5) through (10.14) are valid where the rate of wheelwear is negligible. A more detailed analysis taking into account the continuous wearing of the wheel is given in reference 1.

The maximum stock-removal rate for rough grinding in the reciprocate-grinding mode (as distinguished from creep-feed grinding), can be found by determining the force to stall the wheelhead from Eq. (1.23), the wheel breakdown force F_{bd} from Eq. (1.17), the force to stall the table and the force to dislodge the workpiece from its fixture. The maximum permissible operating force is equal to about 50% of the least of the above forces. The stock-removal rate may be limited by any one of these forces. Once the maximum permissible force is determined, Eq. (10.14) can be used to determine the maximum permissible downfeed d. The limiting wheel breakdown force may have to be reduced somewhat because of the vibration transient that occurs as the wheel strikes the workpiece. It may cause a few chatter marks at the entry edge. They occur close to or at the natural frequency of the system. The resultant force pulsations help to keep the wheel sharp by driving the force intensity F'_n up into the F'_{bd} region (self-sharpening region).

Flatness errors in surface grinding can be predicted by use of the equations presented above and those in Chapter 1. As an example, consider the grinding of a workpiece where the width of contact between wheel and work varies as the wheel passes over the workpiece. The variation in width of contact causes a variation in the "cutting-stiffness" K_c from Eq. (10.11). The variation in K_c causes a variation in the wheel depth-of-cut h, according to Eq. 10.8 so that a flatness error is ground into the workpiece during the rough-grind process. As the wheel is allowed to spark out, that error is reduced and will be completely eliminated after a long sparkout if threshold forces are zero. However, for many difficult-to-grind materials, significant threshold force intensities exist. (See Chapter 1, and Chapter 2) Under those conditions, the flatness error e can be calculated from the difference in deflection caused by the threshold forces at the wide w^1 and narrow w^2 widths-of-cut; thus:

$$e = \frac{F'_{thx}w^1}{K_m} - \frac{F'_{thx}w^2}{K_m} \tag{10.15}$$

Thermal damage and surface integrity problems in surface grinding can often be alleviated by considering the principles presented above. For a given workspeed and width-of-cut, the cutting stiffness K_c is inversely proportional to the WRP which, in turn, indicates the sharpness of the cutting surface of the wheel. Since the sharpness of the wheels may vary as much as 500%, significant changes in K_c result (see Eq. (10.11). As K_c increases because of wheel dulling, the induced normal force F_n also increases according to Eq. (10.14). The threshold of grinding burn and/or thermal damage is exceeded when the wheel sharpness (measured by WRP), and the induced normal force F_n attain certain values. See Chapter 2 for further details on surface integrity. In order to be assured of good surface integrity, it is necessary to monitor and/or control both the wheel sharpness (WRP) and the induced force intensity F'_n and to operate at as high a workspeed (table speed) as possible. See Chapter 14 on Adaptive Control for methods of monitoring and controlling WRP and F'_n.

Reciprocate vs. Creep-Feed Grinding

Although Chapter 12 is dedicated specifically to creep-feed grinding, the principles that distinguish normal reciprocate grinding from creep-feed grinding are presented here. Fig. 10.2 shows 2 distinct regimes: 1, at low workspeed where the stock-removal rate Z_w is directly proportional to workspeed V_w, and 2, where the stock-removal rate is essentially independent of workspeed and in which, as the workspeed is increased, the wheel depth-of-cut decreases, giving the same volumetric stock-removal rate.

In deep cut, creep-feed grinding, the contact area between wheel and work is large, and a relatively large number of grits are cutting at a given instant. The local volumetric stock-removal rate per unit area Z''_w is given by[2]:

$$Z''_w = \text{WRP} \ (\sigma_n - \sigma_{th}) \tag{10.16}$$

where σ_n = normal stress
$\quad\quad \sigma_{th}$ = normal stress threshold below which
$\quad\quad\quad\quad\quad$ no cutting takes place.

In Fig. 10.3, the normal stress σ_n is a maximum at the entering work surface and drops to a low value at the finished exiting work surface.

Since:

$$Z''_w \ (\theta) = V_w \ \sin\theta \tag{10.17}$$

the normal stress can be found from Eq. 10.16; thus:

$$\sigma_n = \frac{V_w \ \sin\theta}{\text{WRP}} + \sigma_{th} \tag{10.18}$$

Also:

$$\sigma_t = \mu\sigma_n \tag{10.19}$$

where μ is the effective coefficient of friction. Then an element of the normal force is:

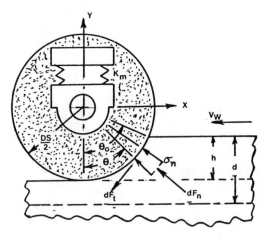

Fig. 10.3 Illustration of creep-feed grinding showing the "unstressed position" d, the wheel depth-of-cut h, and the local normal stress σ_n. (2)

$$dF_n = \sigma_n w \; \frac{D_s}{2} d\theta \qquad\qquad (10.20)$$

$$= \frac{wD_s}{2} \left[\frac{V_w \sin\theta}{WRP} + \sigma_{th} \right] d\theta$$

and the tangential force—

$$dF_t = \frac{\mu wD_s}{2} \left[\frac{V_w \sin\theta}{WRP} + \sigma_{th} \right] d\theta \qquad\qquad (10.21)$$

The vertical and horizontal forces F_y and F_x are found by integrating:

$$dF_y = dF_n \cos\theta \pm dF_t \sin\theta \qquad\qquad (10.22)$$

$$dF_x = dF_n \sin\theta \mp dF_t \cos\theta \qquad\qquad (10.23)$$

between the limits $\theta=0$ and $\theta=\theta_0$ where:

$$\theta_0 = \cos^{-1} \left[1 - \frac{2h}{D_s} \right] \qquad\qquad (10.24)$$

and

Upper sign is used for climb
Lower sign is used for conventional grinding.

The results are:

$$F_y = \frac{wD_s}{2} \left[\frac{V_w}{WRP} \frac{\sin^2\theta_0}{2} + \sigma_{th} \sin\theta_0 \right]$$

$$\pm \frac{\mu wD_s}{2} \left[\frac{V_w}{WRP} \left(\frac{\theta_0}{2} - \frac{\sin\theta_0 \cos\theta_0}{2} \right) - \sigma_{th}(\cos\theta_0 - 1) \right] \qquad (10.25)$$

$$F_x = \frac{wD_s}{2} \left[\frac{V_w}{2WRP} (\theta_0 - \sin\theta_0 \cos_0) + \sigma_{th} (1 - \cos\theta_0) \right]$$

$$\mp \frac{\mu wD_s}{2} \left[\frac{V_w \sin^2\theta_0}{2WRP} + \sigma_{th}\sin\theta_0 \right] \qquad (10.26)$$

Also, the torque on the wheel using Eq. 10.21 is:

$$dT = \frac{D_s}{2} dF_t$$

$$T = \mu w \left(\frac{D_s}{2}\right)^2 \int_0^{\theta_o} \left[\frac{V_w \sin\theta}{WRP} + \sigma_{th}\right] d\theta$$

$$= \frac{\mu w D_s^2}{4} \left[\frac{V_w(1 - \cos\theta_o)}{WRP} + \sigma_{th}\theta_o\right] \tag{10.27}$$

These equations may be simplified when the wheel depth-of-cut h is small compared with the wheel radius; thus:

$$\cos\theta_o \doteq 1 - \frac{2h}{D_s}$$

$$1 - \frac{\theta_o^2}{2!} + \text{---} = 1 - \frac{2h}{D_s}$$

$$\theta_o \doteq 2\sqrt{\frac{h}{D_s}} \quad \text{and} \quad \sin\theta_o \doteq \theta_o \tag{10.28}$$

With these approximations, Eq. (10.25) becomes:

$$F_y = \frac{wV_w h}{WRP} + \sigma_{th}w\sqrt{D_s h} \tag{10.29}$$

When σ_{th} is negligible, the second term may be dropped with the result that the "Cutting Stiffness" K_c is:

$$K_c = \frac{dF_y}{dh} = \frac{wV_w}{WRP} \tag{10.30}$$

The "cutting stiffness" plays an important role in determining whether a grinding operation will be carried out in the "creep-feed" regime or the normal workspeed regime, as shown below.

Eq. (10.29) gives the separating force F_y. It must be in equilibrium with the elastic force generated by the deflection x of the machine and when the threshold stress is negligible:

$$K_m x = K_c h$$

Also, from Fig. 10.3:

$$d = h + x$$

Eliminating x gives:

$$h = \frac{K_m}{K_m + K_c} d$$

The stock-removal rate per unit width Z'_w is:

$$Z'_w = hV_w \tag{10.35}$$

$$Z'_w = \frac{K_m V_w}{K_m + K_c} d$$

When it is recalled from Eq. (10.30) that the Cutting Stiffness K_c is proportional to workspeed, two cases arise:

1. K_c is large compared with K_m
    Equation 10.35 becomes:

$$Z'_w = \frac{V_w K_m d}{K_c} = (WRP) \frac{K_m d}{w}$$

and is independent of workspeed.

2. K_c is small compared with K_m
    Eq. 10.35 becomes:

$$Z'_w = V_w d \tag{10.37}$$

showing that the stock-removal rate is proportional to workspeed.

The criterion for operating in the creep-feed region is:

$$K_m \gtrapprox 5 \, K_c$$

Therefore, creep-feed grinders must be designed to withstand large forces and have a large machine stiffness compared with the cutting stiffness.

The above equations can be used to determine several important characteristics of creep-feed grinding. Eq. (10.18) can be used to calculate the maximum normal stress on the wheel which, in turn, governs the wheelwear rate. Eq. (10.25) or (10.29) can be used to calculate the normal separating force which, in turn, yields the system deflection x during the cut if the system stiffness is known; thus:

$$x = \frac{F_y}{K_m} \tag{10.39}$$

As the wheel runs off the end of the work, that deflection is released, causing a flatness error or exit ramp. Eq. (10.26) gives the force required in the feed direction, and Eq. (10.27) gives the torque and, thus, the wheelhead power required.

CONCLUSION

From the principles presented above, and in Chapters 1 and 2, it is clear that the grinding performance is dependent, primarily, upon the condition of the wheelface as measured by WRP and the normal force F_n. Those 2 important grinding process variables control the stock-removal rate, wheelwear rate, power, system deflection, surface finish, and surface integrity.

References

1.   R.S. HAHN, R. P. LINDAY, "The Influence of Process Variables on Metal Removal, Surface Integrity, Surface Finish and Vibration in Grinding," Proc. 10th MTDR Conf., Univ. of Manchester, Sept. 1969, Pergamon Press.

2.   R.S. HAHN, "On the Universal Process Parameters Governing the Mutual Machining of Workpiece and Wheel applied to the Creep-Feed Grinding Process" *Annals of the C.I.R.P.*, Vol. 33/1/1984, Hallwag Ltd, Berne.

3.   R.S. HAHN, "The Effect of Wheelwork Conformity in Precision Grinding," Trans. *ASME*, Nov. 1955, Vol. 77, 8.

CHAPTER 11

Coated Abrasives

E. J. Duwell

Introduction

The use of abrasive grits on a flexible backing to improve surface finish or shape objects can be traced back to man's earliest attempts to make tools to manufacture useful articles. In recent history, sand, or flint, was held on paper by dried animal glue to make "sandpaper." The term "sandpaper" is still commonly used, even though harder minerals and more durable adhesive and backings have been introduced to extend the product's usefulness. For that reason the term "coated abrasive" is used to broadly include all grit-coated flexible backings.

Coated abrasives are not generally made article by article as are wheels. Rather, wide continuous backings are coated continuously in lengths that may exceed 1 mile and are then stocked in large jumbo rolls. Depending on need, material is removed from the rolls and is "converted" to belts, sheets, and a myriad of other forms (Fig. 11.1).

In this chapter, we will describe the materials with which modern coated abrasives are made and give some guidelines on how to choose the correct product for a particular application. We will also discuss methods of analysis for coated abrasive grinding.

Because many readers may not be acquainted with coated abrasive methods, it may be best to begin by showing some of the many configurations to which coated abrasive tools can be adapted. Belts are certainly the most common form in which coated abrasives are used, but discs, rolls, sheets, and flap wheels also constitute an important market for that kind of abrasive tool. For added information, a reference book is published by the Coated Abrasives Manufacturer's Institute.[1]

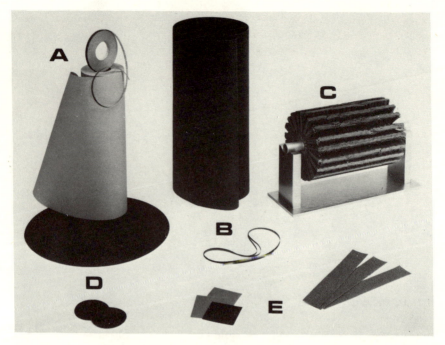

Fig. 11.1   Various converted coated abrasive forms: (a) tapes, (b) wide and narrow belts, (c) flap discs, (d) discs, and (e) sheets

Coated Abrasive Applications

The most common machine for using coated abrasive belts consists of a lathe and backstand (Fig. 11.2) with the belt riding over a contact wheel that may be hard, soft, and either smooth or serrated. Belts may also be run over platens for flat grinding in off-hand applications (Fig. 11.3) or in conjunction with conveyoring belts or tables (Fig. 11.4). Belts are especially useful for centerless abrasive machining or simply centerless finishing (Fig. 11.5). The wide width of a belt and conformability of nonrigid contact wheels makes possible in-line grinding and finishing of cylindrical workpieces in a single pass through multiple heads.

Conformability allows grinding to take place on the slack of a belt, or when using backups such as rollers (Fig. 11.6). Portability is often achieved by using belts on swing frame mounts (Fig. 11.7). Such grinders may be used for weld removal or simple surface conditioning.

For abrasive machining, hard contact wheels carry the belt over the workpiece, which is carried by a rigid support (Fig. 11.8).

Fig. 11.2   Lathe and idler backstand for off-hand belt grinding

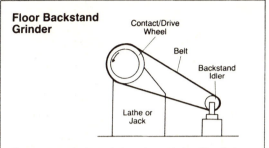

Floor Backstand Grinder

Contact/Drive Wheel

Belt

Backstand Idler

Lathe or Jack

For use on work-pieces that can be carried and handled properly. This machine is basically a conversion of a floor lathe or polishing jack to abrasive belt use by means of a backstand idler pulley. Uses: roughing, blending, finishing, polishing.

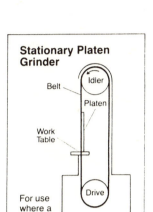

Stationary Platen Grinder

Belt

Idler

Platen

Work Table

Drive

For use where a true, flat surface must be developed. Uses: roughing, blending, finishing, polishing, and precision sizing.

Fig. 11.3   Platen grinder for off-hand grinding

Fig. 11.4   Platen grinders for automated or semiautomated flat grinding: (a) indexing platen and (b) rotary table

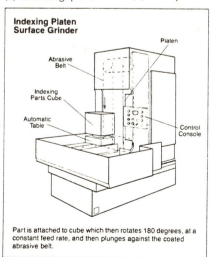

Indexing Platen Surface Grinder

Platen

Abrasive Belt

Indexing Parts Cube

Automatic Table

Control Console

Part is attached to cube which then rotates 180 degrees, at a constant feed rate, and then plunges against the coated abrasive belt.

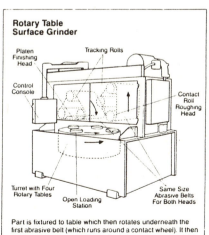

Rotary Table Surface Grinder

Platen Finishing Head

Tracking Rolls

Control Console

Contact Roll Roughing Head

Turret with Four Rotary Tables

Open Loading Station

Same Size Abrasive Belts For Both Heads

Part is fixtured to table which then rotates underneath the first abrasive belt (which runs around a contact wheel). It then rotates under the second abrasive belt which runs around a platen.

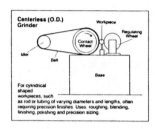
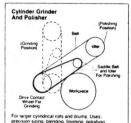
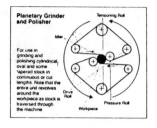

Fig. 11.5 Coated abrasive centerless grinder and cylindrical grinders and polishers: (a) centerless (O.D.) grinder, (b) cylindrical grinder and polisher, and (c) planetary grinder and polisher

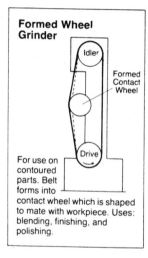

Formed Wheel Grinder

For use on contoured parts. Belt forms into contact wheel which is shaped to mate with workpiece. Uses: blending, finishing, and polishing.

Fig. 11.6 Setup for belt grinding on a contoured idler

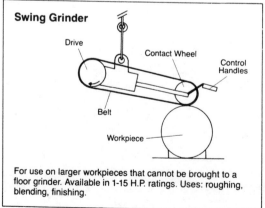

Swing Grinder

For use on larger workpieces that cannot be brought to a floor grinder. Available in 1-15 H.P. ratings. Uses: roughing, blending, finishing.

Fig. 11.7 Swing frame belt grinder

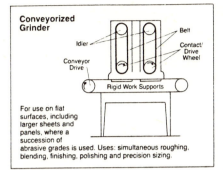

Conveyorized Grinder

Idler

Belt

Contact/ Drive Wheel

Conveyor Drive

Rigid Work Supports

For use on flat surfaces, including larger sheets and panels, where a succession of abrasive grades is used. Uses: simultaneous roughing, blending, finishing, polishing and precision sizing.

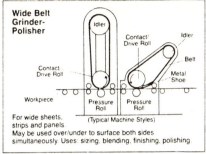

Wide Belt Grinder-Polisher

Idler

Contact/ Drive Roll

Idler

Contact Drive Roll

Belt

Metal Shoe

Workpiece

Pressure Roll   Pressure Roll

For wide sheets, strips and panels. May be used over/under to surface both sides simultaneously. Uses: sizing, blending, finishing, polishing.

(Typical Machine Styles)

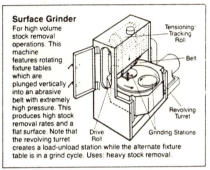

Surface Grinder

For high volume stock removal operations. This machine features rotating fixture tables which are plunged vertically into an abrasive belt with extremely high pressure. This produces high stock removal rates and a flat surface. Note that the revolving turret creates a load-unload station while the alternate fixture table is in a grind cycle. Uses: heavy stock removal.

Tensioning/ Tracking Roll

Belt

Revolving Turret

Drive Roll

Grinding Stations

Fig. 11.8   Belt grinders for continuous flat grinding of sheets and coils or fixtured parts: (a) conveyorized grinder, (b) wide belt grinder-polisher, and (c) surface grinder

Materials of Construction

The essential ingredients for a coated abrasive are a backing, adhesive, and mineral grit (Fig. 11.9). In some products a grinding aid is added as a filler or top coat. The grinding aid serves to prolong belt life by preventing mineral wear or may simply prevent grinding debris from adhering to the troughs between abrasive grits (loading). Grinding aids are particularly beneficial when grinding hard-to-grind metals, such as stainless steel or titanium alloys.

Mineral Types.   Coated abrasives utilize most of the minerals used in grinding wheels. However, since a coated abrasive consists of essentially a single layer of grit on a backing, the cost of the backing and the manufacture of the coated abrasive article results in higher cost per

Fig. 11.9   Cross section schematic of a coated abrasive

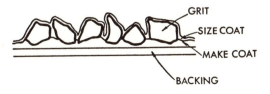

GRIT

SIZE COAT

MAKE COAT

BACKING

pound of mineral than that found in a bonded wheel. To offset that disadvantage, the minerals chosen are usually characterized by high chemical purity, few defects, fracture resistance, and sharp shape. The objective, of course, is to provide a mineral that is worthy of the cost and strength of the backing on which it is placed.

The most commonly used minerals are *flint, emery, and garnet*—these minerals are not sufficiently hard to grind metals and will not be discussed further in this chapter. They are still used to a small extent in wood grinding and some other applications where the workpiece is not particularly hard.

Aluminum Oxide.   Most aluminum oxide abrasive mineral is made by fusion and is modified by only small amounts of titania and some other impurities introduced by the bauxite and not removed in the fusion process. The coating industry tends to use those types of aluminum oxide where the individual grits are single crystals. Thus the weaknesses introduced by grain boundaries and inclusions are minimized.

Aluminum Oxide "Alloys."   Those minerals have been introduced in recent years and are characterized by greatly improved fracture resistance. Two entirely different types are available: an aluminum oxide-zirconium oxide eutectic, which is fusion cast and fast cooled (NorZon)*, and a ceramic oxide made by the dehydration and firing of alumina sols modified with magnesium oxide (Cubitron)**. Those minerals are somewhat softer than the purer aluminum oxides, but their high fracture resistance makes them especially useful for grinding at high material removal rates.

Silicon Carbide.   It is almost always the mineral of choice for grinding ceramics, glass, or cement. However, it is often used in titanium grinding and may be the mineral of choice for grinding aluminum, gray cast iron, copper alloys, and for some stainless finishing processes.

Diamond and Cubic Boron Nitride.   Those minerals are not easily compared with the others listed because of the great disparity of price. They are, however, available as coatings for some applications. For example, the use of diamond or cubic boron nitride coatings on a flexible backing are especially useful for honing, where they will conform to a cylindrical O.D. or I.D. surface of a hardened metal. They are also used

---

*NORZON - Patented by Norton Co.
**CUBITRON - Patented by 3M Co.

on laps and in some belt applications, but those operations will not be discussed in this chapter.

Mineral Grades.   For coated abrasives, the term "grade" refers to grit size and is used in that way throughout the industry. Unfortunately, the term is also used to define the hardness of bonded wheels. The reader is therefore encouraged to interpret the word in the context in which it is used.

Coated abrasives are made in grades ranging from 16 to 600, those numbers corresponding to screen openings per inch used for separating the various sizes. For grinding operations directed at stock removal, grades 24, 36, 40, 50, and 60 are generally used. The finer grades are selected where finish becomes the more important objective.

Adhesives   Abrasive mineral is bonded to coated abrasive backings by animal glue, urea formaldehyde resin, varnishes, and, most commonly, by phenolic resins. That last resin offers the maximum hardness and durability obtainable in the currently made products because of its extreme hardness (Knoop hardness = 24–55) and its outstanding resistance to thermal degradation.

Backings.   Coated abrasives are made on paper, plastic film, vulcanized fiber, and cloth backings. The primary function of the backing is to hold onto the mineral throughout its useful life and to withstand without stretch or tearing the mechanical forces to which it is subjected. Paper or film backings are generally most useful for the surface finishing grades, while cloth backings are used for stock removal grinding where toughness is needed. Water resistance is, of course, also necessary in many applications of that type.

Cloth backings are made of both cotton and synthetic fibers. An integral part of the backing is the saturant, or reinforcing filler placed in the woven structure. The combinations of fiber denier, weave, and filler allow for a broad range of properties that vary from soft and flexible to hard and rigid. The former are obviously chosen where conformance to shaped or soft contact rolls are needed, while the harder rigid types are usually run over very hard metal or rubber wheels.

As manufactured and before converting, the combination of backing, resin bond, and mineral tends to be extremely rigid. In order to impart flexibility to the structure, it is run over 1 or more sharp radii to produce flex cracks in the topmost layer and impart the properties of the backing to the overall structure. The flex cracks separate "islands" of resin and mineral that "hinge" with one another at the flex boundaries. Thus the structure exhibits the properties that are basically supplied by the backing.

Most manufacturers offer a number of "flexes." The most simple consists of a flex pattern that runs perpendicular to the grind direction. The addition or combination of that flex with two flexes at a 45° angle to the grind direction, or a longitudinal flex, results in belts that will conform to a variety of backup situations.

Contact Wheels and Platens

An integral part of coated abrasive grinding processes is the contact wheel, roller, platen, or some other device that supports the belt when a thrust force is applied. To some extent, the effect of those devices parallels the effects obtained by choosing soft or hard wheels (grade).

Smooth steel contact wheels are used for abrasive machining operations where precision and maximum stock removal capability are required. However, an important advantage of coated abrasives is their ability to be "forgiving" in grinding operations where the work dimensions are somewhat inexact or where the work is transferred from head to head. Examples of that are the grinding of wide sheet and coil or centerless grinding on a succession of grinding heads (Fig. 11.10). The use of rubber contact wheels allows for compression of the abrasive belt into the wheel and makes allowance for misalignment of the work or for small part-to-part variations in dimension.

Rubber contact wheels vary in hardness from 90A durometer to 15A durometer. In the range of 70A to 90A, the wheels are used primarily for stock removal. Below hardness values of 60A, the wheels are used for polishing, and at values below 30A, the wheel is expected to deform during operation to "fit the part."

Rubber contact wheels may be smooth or serrated (Fig. 11.11). By changing the land area and the pattern of the serrations, wheels of the same rubber hardness can be made to behave more or less aggressively. Reducing the land area tends to decrease the effective hardness, but it may lead to a more aggressive wheel. The angle of the serrations will also affect the performance of the wheel, with the wheel becoming more aggressive as the land areas approach a perpendicular direction to the direction of grind.

For great conformability, contact wheels have also been made with compressed canvas. On such wheels coated abrasives must be used that have extremely flexible and tough backings.

Softening the contact wheel or platen backup for a coated abrasive always diminishes the total amount of metal that can be removed. For that reason, steel contact drums are almost always used where a combination of high belt efficiency and dimensional accuracy are desired.

Fig. 11.10   Multi-head belt centerless grinder (2 views of the same installation)

Fig. 11.11   Serrated rubber contact wheels for belt grinding

On machines such as the platen grinder shown in Fig. 11.4A, a steel backup is used, but the distribution of thrust force over a large portion of the belt surface results in a substantial decrease in the total amount of metal that can be removed. In machines such as shown in Fig. 11.4C, the platen may be serrated, with the contact surfaces being carbide coated to resist wear under the higher pressures obtained by limiting the contact area in the platen. If the platen portion of the system is only intended to improve surface finish, the platen may be constructed of a softer material and is often covered with glass bead and graphite-coated pad that prevents back wear on the belt.

Stock Removal Capability of Coated Abrasives

Since a coated abrasive consists of essentially a single layer of grit on a backing, the total life of the product depends on the rate at which wear occurs on that single layer of contacting asperities. For the vast majority of applications, the end of a belt's usefulness is reached when the rate of stock removal at some given load becomes too small or if the power requirements for continued grinding at a constant rate exceed the capability of the motor. Metal damage may also define the end of the belt's useful life. In most of those situations, the amount of mineral consumed is only a small percentage of that actually placed on the belt.

Since the cost of an abrasive belt is primarily a function of the area of material in the belt, it is useful to define belt efficiency as a ratio of cut, the total amount of material removed, to path, the area of coated abrasive actually used.

$$\text{Efficiency} = C/P \ (\text{in}^3/\text{in}^2) \qquad \text{(11.1)}$$

The unit for efficiency is simply inches and can be visualized as a quantity of material with an area equivalent to that used in the belt and of a height equal to C/P. The value for C/P of course depends on the product selected, how it is used, and what is defined as an end point in the belt's useful life.

With value given for C/P, the user is able to determine how many parts can be produced with each belt and what the cost of the abrasive belt contributes to the manufacture of the part. At this writing, a ballpark figure for belt cost is about $0.02/in.^{2} of area in a coated abrasive. A belt that is 3 in. × 80 in. in width and length would therefore cost $4.80. If C/P = 1.0, the belt would finish 240 parts if one cubic inch of material has to be removed from each part. The abrasive cost per part would be $0.02.

In order to utilize a belt to its maximum stock removal capability, it is necessary to generate thrust forces that result in metal penetration despite the formation of flank wear support surfaces or cause micro fracture processes that result in the reduction of the flank wear surfaces. The backings and adhesives now used are capable of allowing belt grinding at thrust forces of 500 lbs./(inch of belt width) and cutting forces of 200 lbs./inch (≅40 HP/inch of belt used). Such conditions are not possible on most machines, and therefore the efficiency of coated abrasives tends to be machine limited, especially on old machines. Table 11—1 gives belt efficiencies for grinding plain carbon steel dry under what could be considered light, medium and heavy force levels. Those approximate values are obtained at belt speeds of 5000–8000 ft./min. and will be affected by the use of coolants and lubricants and by the rate of material removal as determined by infeed and throughfeed rates.

For other metals, Table 11—2 shows the approximate cut/path values obtained at the higher thrust and cutting force levels.

The extremely high belt efficiency obtained on cast aluminum and iron components has stimulated the development of machines for coated abrasive surface grinding. Using wide belts and motors in the range of 250 HP, those machines effectively compete with vertical spindle grinders and milling machines in the flat grinding of large parts. A selection of parts is shown in Fig. 11.12. For most of those parts, .035 in. − .125 in. of stock is removed in less than 30 seconds with flatness values of .0015 in. − .002 in. The edges are burr-free.

Table 11—1   Belt performance and grinding parameters for low-med-high degree of severity of belt usage

	Normal Force lb/in Line Contact	Power HP/in	Rate in³/min/in	Cut/Path in³/in²
LOW Sheet Finishing	11 - 33 (2 - 6)	1 - 3 (0.4 - 1.2)	0.33 (5.4)	0.01 - 0.05 (0.025 - 0.13)
MED Off-Hand	33 - 78 (6 - 14)	3 - 7 (1.2 - 2.8)	0.25 - 1 (4 - 16)	0.06 - 0.12 (0.15 - 0.30)
HIGH Pressure Assist	125 - 200 (22 - 36)	10 - 18 (4 - 7)	1 - 8 (16 - 130)	0.1 - 0.2 (0.25 - 51)
	(kg/cm)	(HP/cm)	(cm³/min/cm)	(cm³/cm²)

Table 11—2.   Approximate maximum belt efficiences for various metals

Cast 390 Aluminum	2.0 - 4.0 in³/in²
Cast Iron	2.0 - 3.5
Carbon Steels	0.5 - 1.5
304 Stainless	0.15 - 0.20

Fig. 11.12   Examples of parts machined with coated abrasives. The best parts range in size from 4 in. to 24 in. in swept diameter. Parts with open sections are better than parts with broad flat surfaces. (Courtesy of Timesavers, Inc.)

Dynamics of Grinding with Coated Abrasives

In many applications coated abrasive grinding is done under "constant load" conditions. Examples of that kind of grinding is found in the shaping and finishing of turbine blades or golf club heads on a lathe or in the removal of gates or risers on castings by off-hand grinding. Under constant load conditions the belt initially removes metal very rapidly before it stabilizes and exhibits a more constant behavior during its useful life (Fig. 11.13). (Eventually the development of flank wear surfaces on the contacting asperities results in an unacceptably low rate of cut, a condition often described as "glazing." Also, the specific energy (energy consumed per unit volume of metal removed) reaches a level where metal surface damage may be caused. Power consumption also reaches a very low level.

In actual practice, an operator will tend to increase load as the rate of cut diminishes. That effectively decreases specific energy of grinding, increases the rate of cut, and maintains power consumption at a more constant level. It also results in the regeneration of new sharp asperities by breaking up flank wear surfaces and in the introduction of new cutting grits, which lie at a somewhat lower level above the backing.

For abrasive machining, most coated abrasive grinders are designed to grind at a constant rate of stock removal. Under such conditions, both the load and power curves increase with time, rising sharply when the belt is worn out (Fig. 11.14).

As shown earlier (Table 11—1), the capacity of a coated abrasive to remove stock increases greatly if the mechanics of the process allow for the eventual use of high loads and if the available power will generate

Fig. 11.13 Typical performance trends of dynamic variables for constant load belt grinding

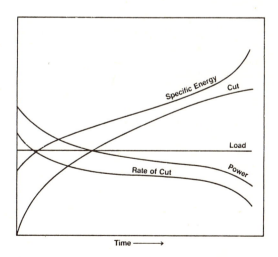

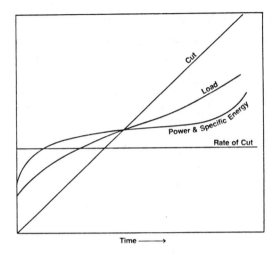

Fig. 11.14 Typical performance trends of dynamic variables for constant rate belt grinding

the needed cutting forces. However, the optimum capacity of the belt to remove stock is greatly influenced by the rate at which the maximum thrust and cutting forces are reached by programming increased loads or by choosing a given rate of stock removal. That is especially true for many hard-to-grind metals. On 304 stainless, for example, increasing the rate of stock removal beyond 1.80 (in³/min)/in. of belt width results in a 50% reduction of belt efficiency (C/P). Thus, the user must choose between increased productivity or improved belt life (Fig. 11.15).

The choice of grinding parameters is generally determined to a large extent by the properties of the metal being ground. A general scheme of relating forces to material removal rates for bonded wheels has been devised[2] that can also be applied to coated abrasives. The normal force, F_N, in constant rate grinding is related to Z_W, the material removal rate by:

$$Z_W = \Lambda_w(F_N - F_{NO}) \qquad (11.2)$$

where F_{NO} is the threshold force where rubbing ceases and cutting begins.

$$\Lambda_w \text{ (metal removal parameters)} = Z_W/(F_N - F_{NO}). \qquad (11.3)$$

The cutting force, F_T relates directly to actual horsepower consumption.

$$HP = \frac{F_T V_S}{33,000} \qquad (11.4)$$

where V_S is belt speed in ft/min and F_T is measured in pounds. Plots of Z_W vs. HP also yield a linear relationship and the determination of spe-

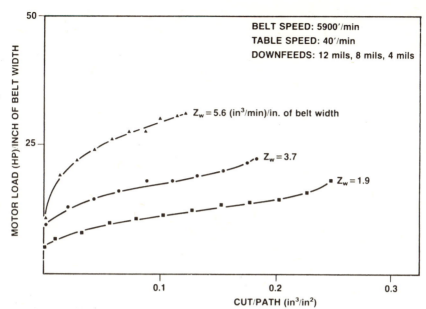

Fig. 11.15   Coated abrasive belt efficiencies for the grinding of 304 stainless steel at rates greater than 1.80 (cu. in./min.)/in. of belt

cific power, P_{SP},

$$P_{SP} = (HP - HP_o)/Z_W \qquad (11.5)$$

where, HP_o is the threshold horsepower consumption generated by rubbing or plowing before material removal begins.

In use, grinding wheels may develop a state of dynamic equilibrium during which the rate of material removal and the rate of wheelwear are constant. The relationships described can therefore be easily applied to both the wheel and the work. For coated abrasives, however, failure of the bond to hold the grit to the backing terminates the belt's life. Since wear of abrasive grits leads to a gradual change in grinding dynamics, the values for Λ_w and P_{SP} must be associated with some parameter that identifies the state of use of the belt. Cut/path (C/P) will serve that purpose. A value is usually chosen that is representative of the belt after "break-in" but before the generation of flank wear surfaces (glazing) causes a sharp rise in the specific energy of grinding.

Plots of Λ_w (in³/min/lb (normal force) and P_{SP} [HP/in³/min)] for a variety of metals are given in Figs. 11.16 and 11.17. The reader will immediately recognize that those metals designed for use under extreme conditions generate low Λ_w values and high P_{SP} values.

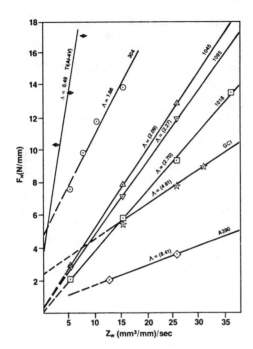

Fig. 11.16 Normal force (F_n) vs. grinding rate (Z_w) for a variety of metals with grade 50 cubitron aluminum oxide abrasive belts

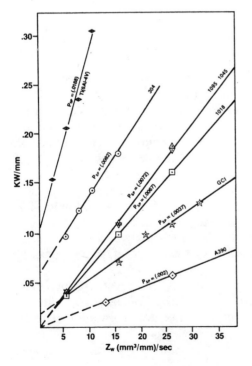

Fig. 11.17 Power (KW/mm) vs. grinding rate (Z_w) for various metals with grade 50 cubitron aluminum oxide abrasive belts

Coated Abrasive Wear and Finish

For any given grade, abrasive wear results in the production of finer finishes as the belt is used. However, surface finish is also improved by the following:

1. decreased grit size
2. softer contact wheels
3. larger land area on contact wheel or smooth wheel
4. higher belt speeds
5. lower throughfeeds
6. higher viscosity lubricants.

For the abrasive machining of gray cast iron at a rate of 2.40 in³/min per inch of belt used, Table 11—3 gives the effect of grades 36–120 on surface finish for throughfeed speeds of 20 ft/min., 40 ft/min., and 80 ft/min. It is apparent that a coarser finish is produced at the higher throughfeed speeds and lower depths of cut. The Ra values are representative of the coarsest finishes produced with those grades. In general, Ra values decrease to at least 50% their original value well before the belt is consumed.

For many multi-head operations, worn belts are moved down the line in mid-life to take advantage of the finer finish produced by the worn belt and to obtain uniform surface finishes. A typical sequence for the finishing of stainless in a 12-head line is shown in Fig. 11.18. When we use 2 grades and finishing with silicon carbide and a high viscosity lubricant, surface finish is reduced from Ra $= 100 \mu$ in. to Ra $\cong 10 \mu$ in.

To avoid deep scratches when both stock removal and finish are required, it is well to choose the finest grade possible on the coarse end that is consistent with grinding rate requirements. In a subsequent series of grinding operations to improve finish, it is generally easier to skip additional coarse grades in favor of using added grades on the finer side.

Table 11—3. Surface roughness values on gray cast iron for various grit sizes and table speeds

		Throughfeed Speed		
		20'/min	40'/min	80'/min
	36	350	395	427
Grit Size	50	295	296	324
	80	167	209	210
	120	81	96	115

$$Z_w = 2.40 \text{ (in}^3\text{/min)in for all experiments}$$

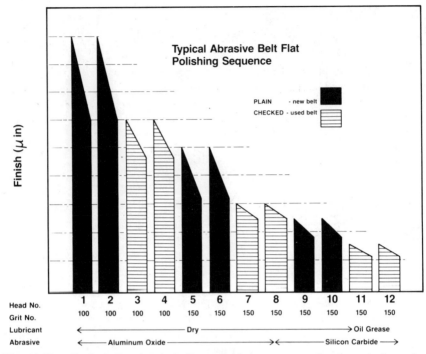

Fig. 11.18   Typical abrasive belt flat polishing sequence for the grinding of a stainless steel. Surface finish is reduced from R_a 90 to R_a 10 in a 12 head operation

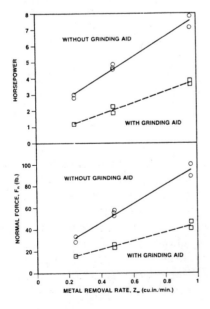

Fig. 11.19   Horsepower and normal force vs. metal removal rate for 304 stainless steel with grade 50 aluminum oxide belts (with and without incorporated grinding aid)

Grinding Aids and Coolants

The grinding process produces conditions that are conducive to the initiation of chemical reactions. Most obvious among those reactions is the pyrotechnical spark shower caused by the oxidation (burning) of the freshly formed chips in air. The freshly formed metal surfaces in the grind interface also apparently tend to reweld to each other and to the grit surfaces, an undesirable phenomenon that is inhibited by air oxidation. It is also restrained by the addition of reactive chemicals to coolants or by the addition of grinding aids to the coated abrasive structure (Fig. 11.19).

Antiwear additives for coated abrasives used in dry grinding usually include sulfur or a halogen. Although the reaction mechanisms of those materials in a grind interface is not well understood, it is usually assumed that they react with the unoxidized metal surface to prevent chip rewelding or adhesion of the metal to the abrasive grit. Their superiority over the influence of atmospheric oxygen may be due to the properties of the reaction products that are lower melting and can be assumed to form lower shear force components in a slide interface. They are especially effective in reducing thrust forces and horsepower in grinding stainless steels and other hard-to-grind alloys (Fig. 11.19).

The array of coolants containing active grinding aids is extremely numerous and complex, and the user must depend on vendor information for suggestions for the best choice for a given metal in a vendor's product line. It can be stated, however, that, as in dry grinding, chemically active lubricants are most beneficial when used to grind the high-performance stainless steel alloys or alloys of titanium, nickel, and cobalt. They are of considerably less importance for the grinding of plain carbon steels (Figs. 11.20a and 11.20b).

Although water is an excellent coolant, it is chemically inert as compared with atmospheric oxygen and will often serve to diminish belt efficiency when not used in conjunction with active chemicals.

Flap Wheels

The conversion of coated abrasives to flaps that are adhesively held in a hub or bundled in a base that can be inserted into a slotted hub results in a tool that can be chosen as an alternative to belt finishing (Figs. 11.21 and 11.22). These abrasive tools are manufactured in a range of grit sizes and, as in the case of belts, can be adapted to the finishing of wide workpieces. Peripheral speeds are generally around 5,500 ft/min., but speeds as high as 10,000 ft/min have been used in descaling applications. Interferences of .060 inches to .120 inches are sufficient to obtain maximum abrasive action from the wheel (interference is the amount of

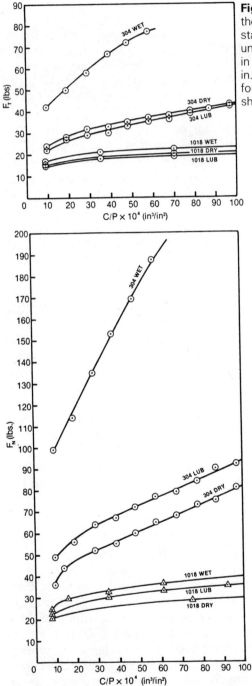

Fig. 11.20   Force vs. cut/path for the grinding of 1018 steel and 304 stainless steel in water, air, and under a flood of sulfur and chlorine in soluble oil (LUB) (Z_w = 1.44 cu. in./min./in. of belt width): (a) normal force shown and (b) cutting force shown

Fig. 11.21 Coated abrasive flap wheel, hub bonded

Fig. 11.22 Coated abrasive flap wheel, bundled segments held in a slotted spindle

downfeed after the wheel first begins to touch the work). For most applications, three quarters to one horsepower are required per inch of wheel width.

For removal of scale, grade 40 silicon carbide is used (oxides are ground better by silicon carbide than aluminum oxide). For reducing surface roughness on clean metals, aluminum oxide in grades as fine as 220–240 is usually recommended. The wheels produce essentially the same surface finish throughout their life.

Because flap wheels can be shaped and are conformable, they are especially useful for cleaning and finishing complex shapes.

References

1. *Coated Abrasives—Modern Tool of Industry*, Coated Abrasives Manufacturers Inst. (CAMI), 1230 Keith Bldg., Cleveland, Ohio 44115

2. *R. S. Hahn and R. P. Lindsay*, "Principles of Grinding," 5-part series in *Machinery Magazine*, New York, July–Nov. 1971, also see Chap. 2, this volume.

CHAPTER 12

Creep-Feed Grinding

Dr. Stuart C. Salmon

Introduction

Creep-feed grinding was developed in Europe in the late 1950s; since then it has been used in an ever-increasing number of applications, particularly in the aerospace industry. Creep-feed grinding in the United States, however, did not enjoy the same success initially, as it appeared that the process was employed using incorrect grinding wheel grades, poor dressing conditions, and less than satisfactory process parameters. The results were not impressive, and after some catastrophies attributable to poor surface integrity, it was assumed to be a high-risk process. Nevertheless, in Europe, the process was a resounding success, and so for many years creep-feed grinding was regarded as "leading edge" technology in the United States. In Europe now the established process is already moving into the new era of Continuous Dress (CD) creep-feed grinding, resulting in enormous productivity gains that have attracted the United States market to look once more at creep-feed grinding both with and without continuous dressing and, in some installations, with the equipment integrated into automated grinding cells and systems producing high-volume high-precision parts.

Creep-feed grinding is generally a surface grinding technique employed for heavy stock removal of even the most-difficult-to-machine materials, particularly where very precise and accurate form is required on the finished surface of the workpiece. There have been applications of cylindrical creep-feed grinding, but they are few.

Fig. 12.1 Reciprocating
grinding process

Conventional Reciprocating Grinding

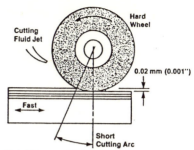

- Low Metal Removal Rates
- Prone to Thermal Damage & Poor Surface Integrity
- Requires Frequent Dressing

Process Comparison

In order to appreciate the impact of creep-feed grinding and be fully aware of situations where it might typically be utilized, it is advantageous to understand and draw a comparison between the reciprocating, creep-feed, and continuous-dress creep-feed grinding processes. (See Figs. 12-1–12.3.)

Before the advent of creep-feed grinding, the typical sequence of operations of manufacture would be first to rough the workpiece by milling, broaching, planning, or hobbing to a size close to the finished dimensions and then, if necessary, to harden the material by some

Creep Feed Grinding

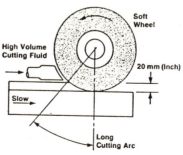

Fig. 12.2 Creep-feed
grinding process

- High Metal Removal Rates
- Prone to Thermal Damage

Continuous Dress
Creep Feed Grinding

Fig. 12.3   Continuous-dress creep-feed grinding process

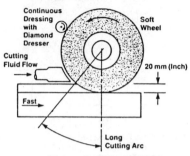

Continuous Dressing with Diamond Dresser

Soft Wheel

Cutting Fluid Flow

20 mm (Inch)

Fast

Long Cutting Arc

• **Highest Metal Removal Rates**
• **Safe from Thermal Damage**

means of heat treatment. The part would probably distort somewhat in heat treatment, but, not too much to prevent reciprocating grinding being used to finish the workpiece. Reciprocating grinding produces a high degree of surface finish and is quite capable of machining the hardened material.

The traditional method described above is inefficient and uneconomic in terms of tooling costs and scrap workpieces. Tooling, in the form of cutters that have to be sharpened, inspected, and stored, are a high-cost item. Cutters are generally inconsistent and often pose problems in areas of cutter jams in broaching and tooth chipping in milling, which can cause scrap workpieces. Heat treatment often causes distortion in small or thin-walled components, which even before grinding will not "clean up" and therefore are already scrap. Reciprocating grinding inherently causes vibration, chatter, and poor surface integrity unless it is carefully monitored. The traditional method also throws up undesirable burrs, which have to be removed, usually by very costly hand benching.

In the case of aerospace alloys the workpiece material is difficult to machine from the outset. Roughing by milling, broaching, etc., is virtually impossible for certain materials and in most cases is not only time consuming but also very expensive. The only economical alternative is to completely grind the material.

The major advantage of creep-feed grinding is that large amounts of stock can be machined from the workpiece, even in the hardened state, with little to no burr. The high cost of deburring processes therefore become an added saving.

Traditionally, for the form grinding of workpieces, the reciprocating grinding process is performed with a multi-station fixture holding many parts. To manually load, clamp, and check the workpiece for position in such a system takes a significant amount of time. That time is unproductive time, as during loading and unloading of the workpieces, the machine cannot run. Dressing is typically carried out using a table-mounted crush roll, diamond roll, diamond block, or even single-point profile dressing. More recently, overhead diamond roll dressing has been employed to dress the grinding wheel owing to the shorter dressing-cycle time. Once the parts have been loaded and the grinding wheel has been dressed, the machining may begin. The machining will have to be monitored continuously for degradation of the grinding wheel. The machining cycle may be stopped many times during the machining of 1 load, so that the grinding wheel may be redressed, owing to loss of form or heavy loading of the grinding wheel periphery, which initiates vibration and chatter, audible to the operator. The initial setup was described as lengthy; however, the cut time, which includes the redressing cycles, is a significantly large proportion of the overall floor-to-floor time. It is a most inefficient process.

The latest advance beyond creep-feed grinding is continuous-dress creep-feed grinding. In contrast with reciprocating grinding, the stock removal rate is extremely high. The cut time is almost insignificant compared with the time taken to load and unload the workpieces. It is therefore essential to automate the material handling in order to reap the full benefit of the faster cut time.

The research carried out at the University of Bristol, in England, showed that by continuously dressing a creep-feed grinding wheel using an overhead diamond roll dresser, the stock removal rate of the process may be increased by a factor of more than 20 times that of conventional creep-feed grinding[3]. In addition to the increase in stock removal rate, an added bonus is that the risk of thermal damage to the workpiece surface is virtually eliminated. The much lower specific energy of the process ensures that a metallurgically sound and high-quality workpiece is produced.

Today in-production facilities are proving the research data, and companies are achieving results that are heralding a new era in abrasive machining. The specialist and aerospace industries are not the only ones to benefit from the new technology. Creep-feed grinding with continuous dressing is competing favorably with broaching, milling, and reciprocating grinding processes as both a high stock removal and high precision surface finishing process.

Reciprocating Form Grinding

Reciprocating surface grinding is the oldest and most popular surface-grinding technique in use in the United States. The design of the reciprocating grinding machine has changed little since its development over the last century. Reciprocating grinding wheels are, in general, hard wheels with little induced porosity in contrast with the newer-technology creep-feed grinding wheels essential for the success of the creep-feed grinding process.

When reciprocating grinding, the grinding wheel depth-of-cut is very small, typically 0.002 in. or so for roughing and 0.0005 in. or so for finishing, depending on the material being ground. The feedrate of the workpiece into the grinding wheel is fast, typically 50 to 80 ft. per minute. The small depths of cut at fast feedrates are an inherent cause of wheel glazing. Glazing occurs when the abrasive grain becomes attritiously worn because of excessive rubbing. The bond strength of the grinding wheel will influence the glazing effect, particularly with workpiece materials of high hardness. A strong bond will hold the abrasive grain in the grinding wheel matrix so that for the grain to break out of the bond, the grinding forces on the grain would have to be sufficiently high to fracture the bond bridges in the wheel matrix. The depth of cut for each individual grain in the reciprocating grinding process is so small that rubbing rather than cutting occurs.

That enhances the formation of flats on the surface of the grinding wheel. The dull grain produces high normal forces and generates enormous amounts of frictional heat. Almost all of the heat generated by the rubbing energy component is conducted into the surface of the workpiece, causing thermal damage and poor surface integrity.

Loading also causes flats on the grinding wheel. Loading is the term given to workpiece material that has adhered to the grinding wheel periphery, causing flats to form. Loading is prevalent in the grinding of soft or "gummy" materials, which cause smearing and redeposition of material on the surface of the workpiece as well as on the periphery of the grinding wheel.

The flats described above, in glazing and loading, often referred to as "wear flats," dramatically increase the normal force on the grinding wheel. The poor rigidity of the reciprocating machine design, coupled with the small depth of cut of the process, can cause chatter and machine vibration to occur, much to the detriment of the surface integrity of the material being ground, the ability to achieve accurate dimensions, and the overall efficiency of the process.

The reciprocating grinding process does not hold accurate forms well. Each local impact of the grinding wheel with the corner of the workpiece breaks away the form dressed on the wheel periphery. A great deal of production time is therefore wasted in redressing the grinding wheel to refurbish the form, particularly as final size is approached.

The efficiency of the reciprocating process is aggravated further by excessive air-cut time at either end of the table stroke. Through hundreds of reciprocating cycles a significant amount of cut time is lost merely in stopping and reversing the table for each pass.

An advantage of reciprocating grinding, however, is that the initial cost of a reciprocating grinder is generally quite low. The process is well-known and accepted by the industry, but so, too, are its shortcomings and low productivity. The growth potential of the process is very poor, in particular with respect to the more difficult-to-machine materials.

Creep-Feed Grinding

Creep-feed grinding is an abrasive machining process, which has a high stock removal capability, particularly for difficult-to-machine materials, and the ability to maintain very close tolerances of both size and form. Creep-feed grinding competes with broaching and milling as both a stock removal and finishing process, producing parts at lower cost per piece and with better-quality surface integrity.

The limitation of the creep-feed grinding process is its inability to machine a long component. In the limit of the creep-feed grinding process the workpiece will incur thermal damage resulting from the degradation of the grinding wheel. That limits the potential length of grind for a given workpiece depth of cut.

The grinding-wheel depth of cut in creep-feed grinding is large, in many cases 10 to 100 times that of reciprocating grinding. Depths-of-cut in excess of an inch are not uncommon in easier-to-grind materials. The workpiece feedrate, however, is slow, comparable to table speeds in milling. The process is analogous to milling but uses a grinding wheel in place of a milling cutter. Typically, the material is removed in 1 or 2 passes of the grinding wheel, initially to rough out the form and then, after a dressing cycle, a finish pass to size. Because the workpiece makes fewer impacts with the grinding wheel, the form dressed on the wheel's periphery is maintained for a longer period of time. That means that the cut time is more efficiently utilized. There is no cross-feed with creep-feed grinding.

Creep-feed grinding is generally performed with vitrified Aluminum Oxide or Silicon Carbide grinding wheels, though superabrasives in vitrified, epoxy, and metallic bonds are being used successfully in more specialized applications. Superabrasive applications tend to be for short workpieces where the full arc length of cut is very much shorter than the length of the workpiece.

The creep-feed grinding process requires a specially designed grinding wheel and a machine tool specifically built for creep-feed grinding.

Machine tool builders in the United States who have recognized the importance and the growth potential of creep-feed grinding have realized that modification of the old reciprocating machine design will not produce the success or the competitiveness required to enter the market with the experienced European creep-feed grinding machine tool builders.

The construction of a creep-feed grinding machine is quite different from that of the conventional reciprocating grinder. Creep-feed grinding exerts much higher forces on the workpiece and requires tight control over the table feed. One basic difference is that the table drive should be a positive mechanical drive as opposed to the hydraulic drives typical of reciprocating grinding machines and also found in earlier creep-feed grinding machines. Indeed, hydraulic drives have been used for creep-feed grinding but only in the up-cutting mode, where the horizontal forces act against the thrust of the hydraulic piston. Down-cutting or climb-cutting is always preferred in creep-feed grinding. Higher spindle horsepowers are required for creep-feed grinding, and hence the machine tools have to be constructed with correspondingly high rigidity to handle the forces and power expected in such grinding. Vibration tends to be less of a problem in creep-feed grinding, as the arc of contact between the grinding wheel and workpiece is very long and tends to dampen any process vibration that might occur. That is not to say that vibration does not occur in creep-feed grinding. Vibration in creep feed occurs from a wheel's being out of balance, stick-slip in the slideway or drive mechanism and dresser transmitted vibration caused by the lack of system rigidity in the dressing unit.

Wheelspeed (RPM) should be a variable in creep-feed grinding in order to keep the grinding wheel surface speed constant across all wheel diameters. Because of the every long arc of cut the chip depth-of-cut for each individual grain must be kept constant.

The key factors to consider when one is using a vitrified creep-feed grinding wheel are a soft grade in the range D-K and a very open structure with moderate skeletal strength. The pore size and distribution must be homogeneous throughout the wheel structure. The high porosity of a creep-feed grinding wheel is essential for the grinding wheel to carry the cutting fluid around the long arc of cut in order to conduct the heat away from the process. The grinding detritus tends to be a very curly and wooley swarf. The large pores in the grinding wheel may also assist in chip clearance. The structure of those wheels is such that by volume, approximately 30 to 40% is abrasive grain, 5 to 10% is bond material, and 50 to 60% is air induced as porosity.

The large depth-of-cut associated with creep-feed grinding creates a long arc of cut. That long arc of cut tends to dampen any process vibration; therefore chatter becomes less of a problem. The long arc of cut means that there are many more abrasive grains in contact with the workpiece than for reciprocating grinding, and though the force on the workpiece is higher in creepfeed, the individual force on each grain is less. Hence the softer grade wheels perform satisfactorily.

Wheel dressing is mostly carried out using a diamond roll when vitrified grinding wheels are used. The object is to dress the grinding wheel as open as possible yet maintain the form accuracy. There are 2 methods for infeeding the dresser. First, there is overhead dressing, where the dressing roll is fed radially into the grinding wheel. Second, there is tangential dressing, where the diamond roll is passed tangentially through the grinding wheel using a table-mounted dressing unit. Overhead dressing is generally the most popular, as the time to dress is faster than that in the table-mounted method. Table-mounted dressing tends to be more accurate; however, the more modern creep-feed grinding machines have closed-loop temperature compensation for spindle growth, and thermal deviations in positioning are negated, as the temperature control keeps the dresser in line with the workpiece fixture at all times.

The correct application of the cutting fluid in creep-feed grinding is most important. The special design of the creep-feed grinding wheel, with its highly induced porosity, helps the situation; however, both high-pressure cutting fluid and high flow rates are essential to ensure that the cutting fluid penetrates into the pores of the grinding wheel and is transported around the arc of cut. Improper application of the cutting fluid is a primary cause of poor results. The cutting fluid has to be high flow rate, in the order of 60 to 80 gpm and at nozzle pressures around 80 to 100 psi. The theory has been postulated that when a jet nozzle is used to apply the cutting fluid, the velocity of the fluid should be equal to or just exceeding that of the peripheral speed of the grinding wheel. That ensures that the cutting fluid penetrates the wheel periphery and is carried through the arc of cut. A common situation that arises on the shop floor is starvation of the cutting fluid toward the end of a cut, resulting in a telltale burning mark. It may be overcome by incorporating a cutting fluid guideway behind the workpiece to maintain the flow of fluid through the arc of cut.

The University of Bristol research showed that there is a significant effect on the threshold of thermal damage in creep-feed grinding from the bulk temperature of the cutting fluid. Experimentation has shown that the cutting fluid warms up as it moves around the arc of cut. The thermal capacity of the cutting fluid decreases with the increase in temperature around the arc of cut. In the limit of the process, the method of heat transfer will change very quickly from convection and conduction, assisted by nucleate boiling to film boiling, which will cause the workpiece to burn. Maintaining a constant, cool cutting fluid temperature will provide thermal control over the process. Refrigeration of the cutting fluid is therefore advantageous. The cutting fluid storage tanks need to be large in order to accommodate a sufficient bulk of cutting fluid to avoid too much churning and energy input to the fluid.

The success of creep-feed grinding in the aerospace and automotive industries prompted the design of the twin wheel creep-feed machines, which were able to grind both sides of a workpiece in one clamping (See Fig. 12.4). In that process the normal forces are canceled out and therefore make the process less sensitive to the high forces that might deflect the part in the fixture. That had a large influence on the design of the future automated grinding cell systems.

Continuous-Dress Creep-Feed Grinding

The combination of creep-feed grinding and continuous diamond roll dressing that resulted from the research work carried out at the University of Bristol was extremely useful. By continuously dressing a creep-feed grinding wheel while grinding, and at the same time compensating for the change in wheel diameter, an enormous increase in stock re-

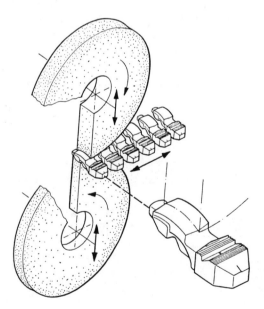

Fig. 12.4   The action of twin wheel grinding

moval rate results. An added bonus is that the risk of thermal damage to the workpiece surface is virtually eliminated.

The study of the creep-feed grinding process centered around the determination of the percentage wear flat area on the periphery of the grinding wheel as a governing factor in the partition of the grinding energy. The total grinding energy for any grinding process is made up of three energies; cutting, ploughing and rubbing. (See Fig. 12.5) Previous research carried out on reciprocating grinding showed that cutting and plowing were a significant amount of the overall grinding energy. Measurements taken and statistically tested during the research have shown the partition of the grinding energy for creep-feed grinding to be quite different.

The key to the success of the continuous dress creep-feed process is the controlled change in the partition of the grinding energy[4]. The total energy of the grinding process is made up of: (1) heating of the workpiece, (2) heating of the grinding wheel, (3) kinetic energy of the chip, (4) heating of the chip, (5) generation of a new surface, and (6) residual stress remaining in the surface of the workpiece. Those energies are generated by the actions of the abrasive, which, depending on their condition, will be rubbing and/or cutting and ploughing. If each of the energies were isolated, then it could be shown that the amount of heat energy conducted into the workpiece surface would be virtually all of the plowing, almost all of the rubbing, and about 5% of the cutting energy. (See Fig. 12.6.) In carefully controlled experiments, the total grinding energy was partitioned and determined to be split 3% cutting and plowing and 97% rubbing, which highlights that the rubbing energy is by far the most detrimental and the chief cause of thermal damage and poor surface integrity. Continuous dressing controls the amount of each of those energies.

Continuous diamond roll dressing allows the accurate control of the

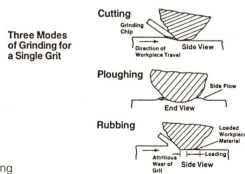

Fig. 12.5 Cutting, ploughing, and rubbing

Grinding Energy

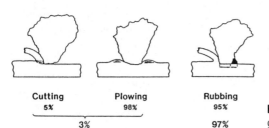

Cutting	Plowing	Rubbing
5%	98%	95%

3%   97%

Fig. 12.6 Partition of grinding energy

grinding process. The surface finish, the forces, the stock removal rate, and the form retention are all influenced by the action of continuous dressing. Continuous dressing negates the hardness of the grinding wheel and allows an increase in stock removal rate of 100 times for the standard wheel and 25 times increase for the creep-feed grinding wheel. (See Fig. 12.7.)

The research carried out using the continuous dressing process allowed strict control of the grinding parameters; hence the relationship between the percentage wear-flat area and the specific energy of the process could be accurately determined. The continuous dressing action eliminates the majority of the rubbing energy component and allows the grinding wheel to cut more efficiently. With continuous dressing, the grinding energy is directed more toward shearing the material and leaves the work zone in the form of heat and residual stress in the chip. The grinding forces on the workpiece drop dramatically.

Continuous-dress creep-feed grinding allows a dramatic increase in the stock removal rate by keeping the grinding wheel sharp and at the

Material	Wheel Grade	Creep-Feed Grinding		With Continuous Dressing	
		Feed Rate (in./min.)	Limitation	Feed Rate (in./min.)	Limitation
C 1023	WA 60 KV	0.400	Burn	41.75	Burn
MAR M002		0.400	Burn	46.06	Burn
C 1023	WA 6080F P2V	2.362	Burn	41.73	Wheel Breakdown
MAR M002		2.165	Burn	53.15	Wheel Breakdown

Depth of Cut — 0.120"
Wheel Dia. — 24"

Fig. 12.7 Effect of continuous dressing on feed rate

same time maintaining the form on the wheel periphery. If one attempts to increase the stock removal rate above a certain limit, the wheel break-down faster than the continuous dress feed rate and contact with the dresser is lost. That protects the workpiece from thermal damage. The workpiece is machined oversize so that there is theoretically zero scrap. The part may be remachined with the confidence that the surface is free from metallurgical damage.

That is particularly true with respect to continuous dressing, yet it may also apply to all forms of diamond roll dressing; for the lowest spe-cific energy and therefore the highest stock removal rate, the grinding wheel must be conditioned to be as sharp as possible. The diamond roll should therefore operate at peripheral speeds approaching the grinding wheelspeed, ideally at .8 of the peripheral grinding wheelspeed. Speeds for diamond rolls 4 to 6 in. in diameter are typically in the range of 3000 to 4500 rpm. A closely controlled radial infeed of the diamond roll, de-pendent on the grinding wheel diameter, is essential, too. The max-imum radial dresser infeed rate is approximately 0.00008 in. per revolution of the grinding wheel. That is an exceptionally high dresser infeed rate and can generate heavy wear on the diamond roll periphery. Typically the dresser infeed rate is in the range 0.00001 in. to 0.00004 in. per revolution of the grinding wheel. (See Fig. 12.8.) That will depend on the material, the depth-of-cut, and length-of-cut relative to the time taken to load and unload the workpiece. The economics of the process are dictated by the wheel-wear rate that affects the machining time of the workpiece. It becomes critical to balance the machine time with the load/unload time, especially in automated systems.

It is essential that the dressing system remains dynamically stable with respect to the operating rpm and the diamond roll form width being used. The system must also be sufficiently rigid to feed accurate increments of the dresser infeed against the dressing forces.

Continuous Diamond Roll Dressing

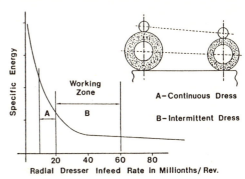

Fig. 12.8 Relationship between the specific energy and dresser infeed rate

Continuous-dress creep-feed grinding gives the impression of high grinding wheel usage, but that is not necessarily so. The process uses little to no more grinding wheel per piece than the creep-feed grinding process. Wheel changes may have to be made more often because the production rate is an order of magnitude higher than creep-feed grinding.

The increase in productivity has changed the face of abrasive machining in that the cut time is now a very small part of the overall floor-to-floor time. Though the specific energy of the process is small, the machining time is short and therefore demands much higher peak horsepower from the machine spindle. Today's continuous dress creep-feed grinding machines are equipped with spindle drives between 45 and 100 hp. Consequently the machine must be rigid and both statically and dynamically stable. Some machine tool builders offer an epoxy concrete machine base, which increases the overall stability of the machine tool. The abrasive machining system should feature automatic operation, automatic parts handling, automatic inspection and feedback, and automatic wheel changing with on-machine balancing for fast setup. All are features in a modern grinding machine design essential to achieving the full potential from the continuous-dress creep-feed grinding process.

Process Summary

It should be understood that creep-feed grinding and creep-feed grinding with continuous dressing are not the panacea for all abrasive machining applications. Reciprocating grinding may be cost effective for form-grinding easy-to-grind materials where there is little stock to be removed and the form to be ground is not too accurate or complex. Creep-feed grinding is suited for short parts with a moderate amount of stock to be removed in difficult-to-machine materials. Continuous-dress creep-feed grinding will machine large volumes of stock very quickly, maintaining very accurate form profiles, independent of the length of the workpiece. Creep-feed and continuous-dress creep-feed grinding are therefore taking the place of milling and broaching operations.

The advantages of creep-feed grinding are high productivity, the ability to machine difficult alloys, the elimination of milling processes, more accurate form retention, and excellent surface finish. Creep-feed grinding is a more efficient process than reciprocating grinding in that the machine table movement does not greatly overrun the workpiece; hence a greater portion of the floor-to-floor time is spent machining. The main disadvantage of intermittently dressed creep-feed grinding is that thermal damage will occur, resulting in poor surface integrity and metallurgical damage. In sharp contrast, *continuously dressed* creep-feed grinding machines are able to operate at very fast rates and can be controlled to eliminate the risk of thermal damage to the workpiece.

Advanced Abrasive Machine Tool Concepts

The continuous-dress creep-feed grinding process has caused a revolution in machine tool design[5]. The highly productive nature of the process virtually eliminates manual operation so that the basic construction of the machine tool can incorporate a more rigid structural system. Early on in the research it was realized that a production machine would require a great deal more power than conventional creep-feed grinders do. Machines suitable for creep feed with continuous dressing should be rated around 45 to 100 hp at a spindle rpm in the range of 4500 to 6000 sfm wheelspeed.

The continuous dressing system requires close control to maintain the 0.8 wheel-to-dresser, synchronous speed ratio. The bearing and infeed mechanism of the diamond roll unit are susceptible to machine vibration with respect to the stable width of the dressed profile and thermal cycling. The dressing system must be exceptionally rigid and thermally stable for the success of the process. The dresser drive should be a direct electric motor drive and the infeed drive should be a direct, typically ball-screw, mechanical drive with feedback positional control. The dressing system is the most critical area of the machine tool, as it not only controls the continuous dress capability but also governs the accuracy of the entire process.

The cutting fluid should be supplied in a copious manner and should be refrigerated for both dimensional and process consistency. An important part of the cutting fluid system is the need to pay close attention to the filtration and disposal of the grinding detritus, which floats on the surface of the fluid, making purely cyclonic or weir-type cleaning systems unsuitable and unreliable. The continuous-dress process will generate grinding swarf in vast quantities. The need for a system to deal with large volumes of "woolly" swarf is paramount. A traveling band-paper filter, inverting bag, or skimmer system is recommended with cyclonic fine filtering as backup. Depending on the system chosen, the bulk quantity of the cutting fluid in the total system will have to be calculated carefully to give time for adequate filtration and thermal stabilization.

Grinding wheel grades are less critical with continuous dressing because of the negation of the effect of wheel hardness; however, the type of abrasive grain may change. Aluminum Oxide wheels do not machine titanium at all well, whereas Silicon Carbide machines titanium extremely well. Both CBN and diamond have applications, too, for different materials, part configurations, and batch sizes, and they can be used to advantage on the newer creep-feed machines because of their increased power and rigidity.

More complex grinding machine tool controls are available which, even when continuously dressing, can raise and lower the wheelhead as the machine table traverses. That interpolation control feature allows for contouring along the machine table axis.

In order to reach a high level of flexibility, the machine tool should have provision for the automatic changing of the grinding wheel and on-machine balancing. A means of automatic dresser changing should be incorporated within a system that combines sensing and control feedback so that the machine is capable of automatic setup up with respect to given dimensional data.

A similar detection and sensing system discussed above should be incorporated on the machine table. A conventional part-fixturing system bolted to the machine table will detract from a flexible cell system. Ideally a universal fixture, capable of clamping all shapes of workpieces, would be ideal. Such a fixture may be possible depending on the configuration of the workpieces to be machined in the cell.

The application of the process has centered around the aerospace industry, in particular, the manufacture of turbine blades, which were already machined by an abrasive machining process. There are now applications in the field where creep feed and creep feed with continuous dressing are replacing milling and broaching operations, even on a job shop scale. The systems have to be designed with high flexibility because of the tremendous productivity, particularly for small job shop batch sizes. The manufacture of gears, pump rotors, and cutting tools is being revolutionized. Cylindrical applications have less impact owing to the high-speed turning capability of CBN and polycrystalline diamond turning tools. Applications for cylindrical creep-feed continuous-dress grinding may arise in difficult-to-machine materials, which are sufficiently out of balance to prohibit high-speed turning, making room for a slow rotational creep-feed grinding operation. Applications for tooling manufacture are on the increase: Roughing and complete finishing of broaches, machining hobs, saw blades and thread chasing dies in the hardened state, from the solid, etc.

The abrasive industry has reached a point where the key to its future success is the combination of the right level of automation. The process of reciprocating grinding—which traditionally, as a proportion of the overall floor-to-floor time, has taken more time with respect to the part load-unload time, dressing, and wheel and dresser changing, among other things—is now a process where the machine cut time is a minimal amount of the overall floor-to-floor time and the part load-unload time, and so on have now become the limiting factor.

Grinding Cells - Factories of Flexible Automation

The continuous dressing process realizes enormous reductions in machining time per piece. Typically a turbine blade that by creep-feed grinding would take minutes to grind now requires only a few seconds to machine.

Though not ideal, existing creep-feed grinding machines may be modified to accommodate the continuous-dress creep-feed process and achieve reasonable results. However, tread with the utmost caution should machine modification be your aim. Early in 1980 Rolls-Royce Limited in Derby, England, decided to try the modified route, forcing a lead in the manufacturing technology. With a large creep-feed production facility, the existing machine designs were modified to operate with overhead continuous dressing and automatic robot loading and unloading of the parts to be machined. A cell comprising 7 grinders with robot loading/unloading, can perform 1 grind operation on each machine[6]. The turbine blades are then cleaned and inspected before handing off to the next machine for subsequent operations. (See Fig. 12.9.) The blades were held and transported in a low-melting-point zinc alloy matrix.

A disadvantage of the 7-cell line at Rolls-Royce was the multiple clamping of the zinc matrix block. The heavy handling of the matrix block creates inaccuracies in location and movement of the turbine blade within the matrix owing to the clamping forces.

The second generation of abrasive machining cell for Rolls-Royce is now in operation[7]. It consists of a special purpose-built machine, capable of machining multiple operations in a single setup. The encapsulated blade is automatically loaded into the machine tool, which has 2 horizontal spindles 1 above the other. The upper spindle has 3 ganged wheels and the lower spindle 2 ganged wheels, allowing 5 surfaces to be machined in 1 loading. One blade is manipulated and machined at a time. That machine tool concept will allow a shrouded turbine blade to be completely machined in a minimal number of clampings. (See Fig. 12.10.)

In 1978 a program RAMIGO (Robotics and Automated Measuring in Grinding Operations) for grinding turbine blades began in Norway. The Norwegian Company of Kongsberg, in a supporting role for European

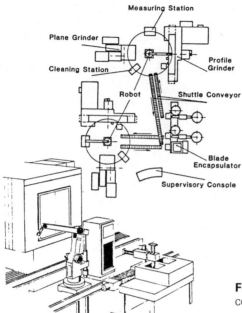

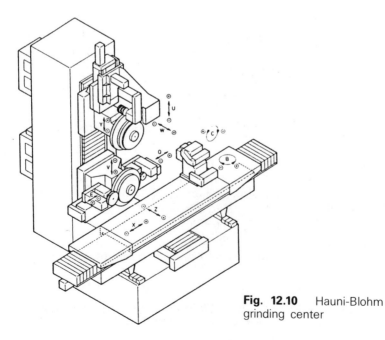

Fig. 12.9   Grinding cell concept at RR.

Fig. 12.10   Hauni-Blohm grinding center

defense, contracted to machine turbine blades, using creed feed only, for the F16 aircraft. Later in 1982, after an extensive feasibility study, the company was assisted financially by the USAF and Pratt & Whitney Aircraft to complete the project. The RAMIGO system ran production hardware in January 1984 and showed how an automated grinding cell system can be integrated and function both profitably and successfully[8]. The system has realized a 50% increase in capacity along with a 75% reduction in manpower.

Control of such a factory system is complex. It must be a 2-tier system, the lower level at machine level, where the control monitors the machine status, facilitates diagnostics of the machine system, and controls the parts-handling system and inspection feedback loop. The upper tier is that of factory management, tracking parts through the system, automatically scheduling the machines in the cell, and monitoring performance. The overall key to success is flexibility in such a cell. Once a family of parts has been defined, then the system should be designed to cope

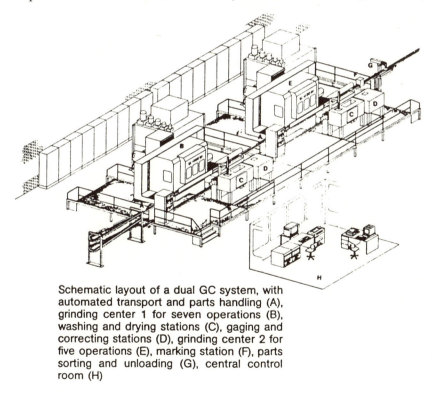

Schematic layout of a dual GC system, with automated transport and parts handling (A), grinding center 1 for seven operations (B), washing and drying stations (C), gaging and correcting stations (D), grinding center 2 for five operations (E), marking station (F), parts sorting and unloading (G), central control room (H)

Fig. 12.11   Schematic layout of the automated grinding center system

with that family and a growth family in order to minimize changeover times, setup, and complement readiness, and surge capability.

Today grinding cells are operational in Rolls-Royce, England, and Kongsberg, Norway, where machines are being loaded by robots. In 1984 Rolls-Royce took delivery of a more complex system, which has the 2-tier control, machine diagnostics, statistical quality control, and a management supervisory network. Two machines complete a shrouded turbine blade in only three machine clampings. (See Fig. 12.11.)

The advent of Continuous Dressing coupled with creep-feed grinding is heralding a new era in abrasive machining. It has been recognized that in order to obtain the full potential from the process, automation is the key. Unfortunately, those new principles of advanced abrasive machining require considerable corporate investment. The technology is spreading fast, both U.S. Machine Tool Builders and users have realized that the available quantum leap in productivity is a formidable force to compete with. Abrasive machining has seen little advances since its early beginnings, but all at once the sparks are flying[9].

References

1.   Jablonowski J., "Will Creep-Feed Grinding Catch On?" *American Machinist*, December 1980.

2.   Zhou Q.Z., "Cylindrical Creep-Feed Grinding," Shanghai Machine Tool Works, People's Republic of China.

3.   Salmon, S.C., "Creep-Feed Surface Grinding," Ph.D Thesis, University of Bristol, England. 1979.

4.   Salmon, S.C., "Advanced Abrasive Machining," Annual Conference, Abrasive Engineering Society, May 1982.

5.   Albert, M., "Taking the Creep Out of Creep-Feed Grinding," Modern Machine Shop, November 1982.

6.   Rolls-Royce Automating Blade Line, *Aviation Week,* December 6, 1982.

7.   Radford, W.F., & Redeker, W., "Flexible Automation for Profile Grinding," Werkstatt und Betrieb, 1983.

8.   Manty, B.A., et al., "Robotics and Automated Measuring in Grinding Operations (RAMIGO), 15th SAMPE Conf., Cincinnati, Ohio, October 1983.

9.   Salmon, S.C., "Creep-Feed Grinding with Continuous Dressing—A New Era, SME Intl. Grinding Conf., MR84–539 August 1984.

CHAPTER 13

HONING

Hans Fischer
Sunnen Products Co.

Introduction

What Honing Is

Honing is an abrasive machining method, most often used to improve the accuracy of internal cylindrical surfaces, and it is characterized by (1) a large area of abrasive contact, (2) a low cutting pressure, (3) low velocity, (4) a floating part or floating tool, and (5) automatic centering of the tool by expansion inside the bore.

Reasons for Honing

Honing can be used to improve the geometry of a hole that has been created by a method of lesser accuracy, such as casting, punching, drilling, reaming, boring, or even grinding. Honing will follow the original centerline of the hole; therefore it can *not* be used to change the location of a hole. That is an important fact to remember when one is scheduling the sequence of manufacturing operations. Normally the amount of stock removal is small enough and the tolerance for the concentricity between the inside and outside diameters is large enough to take care of the problem. But when the tolerance is close, it may be necessary to hone the hole, and then machine the outside of the part with respect to the honed hole.

Fig. 13.1 both shows 10 bore errors that are caused by machining, chucking, or heat treatment. Honing can remove all those errors, and it can do so with the least possible stock removal.

301

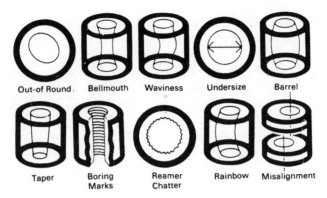

Fig. 13.1 Ten bore errors

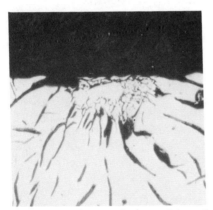

Fig. 13.2 Sectional view of bored automotive cylinder (x100)

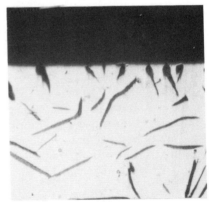

Fig. 13.3 Sectional view of bored automotive cylinder after honing (x100)

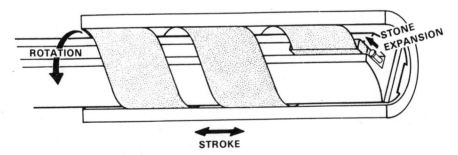

Fig. 13.4 Basic process of honing

Honing is gentle to the material being honed. The surface integrity of metal after honing is excellent, while more abusive machining methods, such as grinding or boring, may fracture the crystals of metal to a depth of about .002 in. (0.05mm). That is why, for instance, a bored engine cylinder is honed, to establish not only good dimensional accuracy (diameter, straightness, and roundness) but also a sound, base metal surface.

Application

Basic Concepts

Honing applies 3 forces:

1. *Stone expansion* forces abrasive against the wall of the workpiece.
2. *Rotation* of the honing tool (or workpiece), combined with
3. *Stroking*, removes material, creates characteristic crosshatch pattern. See page 324 for more infomation about crosshatch.

Honing can remove the errors caused by machining methods of lesser accuracy because:

(1.) No chucking is required. Small parts are supported by the honing tool.

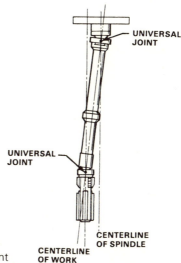

Fig. 13.5   Universal joint

(2.) No alignment is required. The part floats and aligns itself automatically on the expanding tool. If the part is heavy, the tool
floats on 2 universal joints and aligns itself by expanding inside
the part. (See Fig. 13.5)
A honing tool does not depend on bearings or supports outside
the hole being honed. It supports itself inside the very hole it is
honing. It is not affected by the quality or the age of the bearings
of the honing machine.

(3.) Honing tools are rigid. That removes out-of-round, bellmouth,
waviness, and barrel.

(4.) Honing tools have long abrasive sticks in contact with the bore
unlike the line contact of a grinding wheel. That helps remove
waviness, taper, rainbow, and misalignment.

(5.) Honing tools have stones and guides arranged at the proper angles to avoid chatter. That fights out-of-roundness and reamer
chatter.

(6.) The length of the hole to be honed is practically unlimited. Standard tools can hone a hole .060 inches diameter, 1.25 inch long
(1.52 × 31.75mm), which is a length-to-diameter ratio of 23; or 1
inch diameter, 18.5 inches long (25.4 × 469.9mm), which is a ratio
of 18.5; or 2.9 inches diameter, 120 inches long (73.6 × 3048mm),
which is 41 times as long as the diameter. Those ratios are impos-
sibly long for a tool supported by an outside bearing, such as a
grinding wheel or a boring bar.

Tool design.   Fig. 13.6 shows a typical honing tool. Note the great
stone length and the rigidity made possible by supporting a row of relatively short stones by a common wedge and the fact that only the row of
stones moves, while the guide shoes are stationary.

An end view of the same honing tool shows the unevenly spaced 3-

Fig. 13.6   Typical honing tool

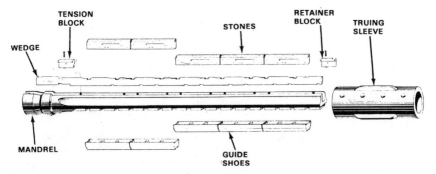

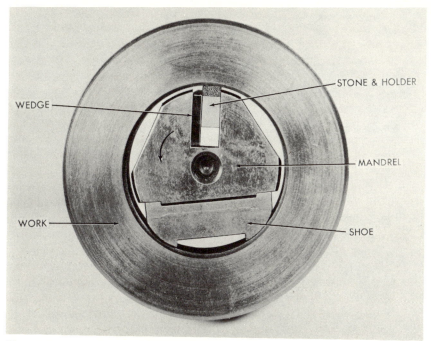

Fig. 13.7   End view of honing tool

line contact, which stabilizes the tool and creates roundness and straightness with the least possible stock removal.

That selective stock removal means that the big areas of an inaccurate hole will be touched by the honing stone only after the tight areas have been honed as large as the big areas. Such "cleaning up" action can be seen by looking into a partially honed hole and observing the difference between the honed and unhoned areas.

The design of the tool is very important: Tool rigidity avoids following an out-of-round or nonstraight condition.

If the contact lines were evenly spaced, they would produce a rhythmical motion, called chatter, actually *generating* a lobed (out-of-round)

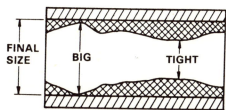

Fig. 13.8   Selective stock removal

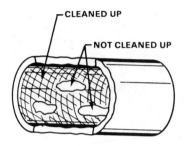

Fig. 13.9   Partial clean up

hole. If the angular difference between the 3 contact lines were improper, it would also chatter, but at a higher frequency, creating objectionable noise, a corduroy pattern finish (rougher than normal), and poor stone life.

Abrasive.   Aluminum oxide and silicon carbide are the conventional abrasives. Borazon (Cubic Boron Nitride, or CBN) and diamond are the superabrasives. Different materials respond better to 1 type of abrasive grit than to another type. Aluminum oxide is best suited for stock removal in steel. Silicon carbide works best on cast iron, bronze, brass, aluminum, and for fine finishing steel and on some nonmetallic materials, such as acrylic, Dapon, Delrin, epoxy, graphite, Kel-F, Lucite, nylon, phenolic, Plexigrass, polycarbonate, polyethelene, Teflon, Torlon, Zytel, etc. Borazon is useful for honing any kind of steel, hard or soft, high or low alloy, when used in a metal bond. Bonds are further discussed below. Diamond is the only abrasive that works for tungsten carbide, glass, and ceramics, and it is also effective in cast iron. Most honing is done with conventional abrasives because of their low price. However, the use of superabrasives is increasing rapidly, because high abrasive *price* does not necessarily mean high *cost* of honing. A comparison of costs is given below.

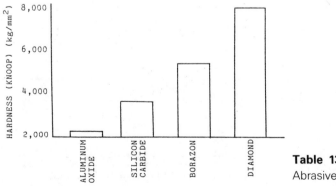

Table 13—1
Abrasive hardness

Fig. 13.10   Bonded stick for abrasive grains

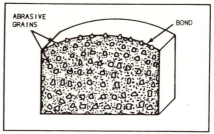

Bond. The bond holds the individual grits of abrasive together. Vitrified bond is most commonly used for conventional abrasive. A type of clay is mixed with the abrasive grits and then fired in a furnace to a glasslike (vitreous) consistency. Resin bond used to be shellac but now is usually plastic. Its use in honing is limited to very fine polishing or to Borazon stones for use in smooth holes in hard steel and honing extremely hard and accurate parts, such as ring gages or fuel injector barrels.

Metal bonds can be brass, steel, nickel, or carbide based. Metal can be used to make bonded sticks or plated surfaces. A bonded stick is composed of bond matrix and the abrasive grains. It is designed so that, as the top layer of abrasive grains wears, they break out and expose sharp new grains further down in the stick. Plated products consist of a single layer of abrasive grains attached to the surface by electrolytic plating. Fig. 13.10 and 13.11 should clarify the differences between the 2 bonding systems. The single-layer plated surface is used for Single-Stroke Honing tools. More about single-stroke honing tools below.

The effective hardness of a honing stone is determined by the type and quantity of bond as well as by the closeness of the grain spacing (density) and, or course, the abrasion resistance of the type of abrasive grit.

The abrasion resistance of each type of abrasive grit is not always pro-

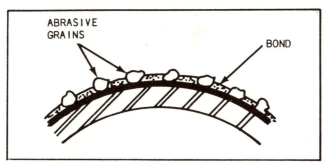

Fig. 13.11   Plated surface for abrasive grains

Fig. 13.12a Hand workholding fixture

portional to its hardness. For instance, diamond, the hardest of all materials, might be assumed to be good for honing steel. However, diamond does not always perform well on steel. Theory has it that because of steel's affinity for carbon, the diamond, which is carbon, is actually dissolved into the carbon-poor steel. While that theory holds for all applications of diamond in steel, the bond being used to hold the abrasive must also be considered before making a final judgment about the appropriateness of the particular abrasive. For instance, because of superior strength, diamond *is* used on plated single-stroke honing tools for honing steel but is *not* used in bonded abrasive sticks for conventional multi-stroke honing of steel.

Hardness of a honing stone is not a measure of quality. The best stone for a job is one hard (strong) enough to hold each abrasive grit just long enough to use up its sharp cutting edge. A honing stone must allow the dulled grits to drop out and permit the sharp grits underneath to take over.

Machine Considerations

Both horizontal and vertical honing machines are available, but no proof is available that 1 type gets better results than the other in general honing, neither is speed nor accuracy attainable. There are some obvious limitations. For instance, to hone a 10-ft.-long (about 3 meters) tube, a

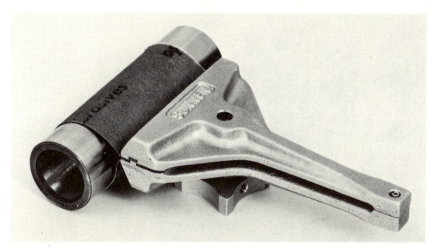

Fig. 13.12b Hand workholding fixture

Fig. 13.13   Workholding fixture for power stroking

This is an example of how to fixture a heavy part. The tool floats while the part is locked in place.

Swings out for rapid Unloading and Loading

Accurate positioning of part is not required.

Part ready for Loading or Unloading

The part is end clamped, which is less likely to distort than chucking on the out-side diameter. It is also safer. The part can't jump out of the fixture, even if loose nuts may permit it to spin in the fixture.

CRADLE   *Part ready for Honing*

Fig. 13.14   End clamping fixture

vertical machine would have to be at least 26 feet (about 8 meters) tall, and it would be difficult to find a building to fit it in. On the other hand, short heavy parts, for example, with a 6-in. (about 150mm) hole, 10 in. (about 250mm) long, would make it easier for the operator to have a vertical machine, because he would not have to lift the tool in and out of the hole and support the tool's weight.

Honing machines are available, ranging from the simplest, basic machine to automated, robotized versions, selling for a hundred times as much. There are 4 common types of honing machines. The basic hone requires the operator to hold the part in his or her hand while stroking the workpiece back and forth over the honing tool. Another is the power-stroked machine. The operator puts the part into a fixture, starts the spindle, stops, and unloads the finished part. Still another is the automatic honing machine, which gages the hole diameter while it is being honed and stops the honing operation when size has been reached. And the most expensive machines will do *everything*: loading, gaging, unloading; some will even segregate the parts according to minute differences in size, label them, and present management with a printout of statistical quality control.

Fixturing

Fixtures for honing are usually quite simple, because it is not necessary to locate the workpiece accurately, and the forces used to hone are usually quite low. A lot of honing is still done by hand. The operator moves the part back and forth over the honing tool by hand while holding the workpiece in a loop of emery cloth, held in a clamp to absorb the torque of honing. This same fixture is used for power stroking. This is an example of how to fixture a heavy part. The tool floats while the part is locked in place. Accurate positioning of part is not required. The part is end clamped, which is less likely to distort than chucking on the outside diameter. It is also safer. The part cannot jump out of the fixture, even if loose nuts may permit it to spin in the fixture.

External Honing

The great majority of all honing is done to interior surfaces. External surfaces do not usually have the accuracy problems that internal surfaces have. It is easier to accurately machine outside surfaces because the tool does not have the size limitations imposed by having to fit into the hole. But even so, honing may be used to overcome accuracy problems that can result from grinding. This ground hardened steel plunger looks very good. But, after the surface has been blackened and a light pass taken with a fine grit hone, the surface imperfections become

Fig. 13.15   Steel honing plunger, before use

Fig. 13.16   Steel honing plunger, after light pass

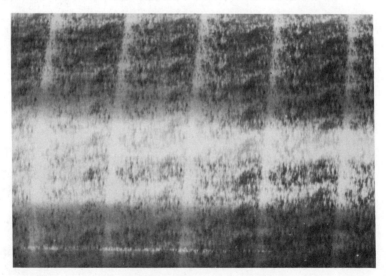

Fig. 13.17   Steel honing plunger, after heavy-chatter grinding operation

Fig. 13.18   Plunger after honing

Fig. 13.19   Hand honing operation

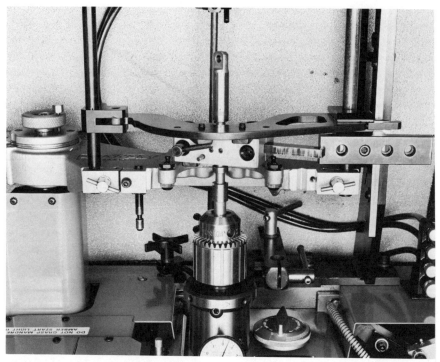

Fig. 13.20   Power honing

clearly visible. A magnified view of the same surface shows the chatter marks of grinding, resulting in rough finish, out-of-roundness, and undefinable size. External honing has removed the imperfections; the part is accurate within .0001 inch (0.0025mm). A skilled operator with proper gaging equipment can get as close as .000010 inch (0.25 micron). External honing can be done by hand . . . . . . or by power stroking.

Honing Oil

Commonly the fluid used for honing is referred to as a "coolant." That is a misnomer, because cooling is not a strong point of any oil. By far the most important reason for using oil for honing is its chemical activity. Good honing oil must be inactive at normal temperatures, so it does not corrode anything. But it must instantly become active when the temperature comes close to the melting point of the metal being honed. That high temperature occurs in microscopic spots at the point of cutting action and would result in welding of the metal guide shoe to the metal being honed. Those tiny weld spots would be torn apart by the force of the honing machine, and the results would be rough surface finish and rapid wear of honing stone and guide shoe. However, capable honing oil will prevent welding by chemically changing the hot spots from metal to a nonmetallic compound, which cannot be welded. That welding problem is especially likely to happen with high-alloy materials, such as stainless steel.

Honing tools without guide shoes are not immune to this welding problem. If the honing fluid does not have enough chemical ability, it will permit stone loading, which means that the metal chips created by the cutting action of the honing stones can stick to the stone surface. Then there will be metal-to-metal contact and welding, with the same undesirable results as when one uses a honing tool with metal guides.

Getting Ready to Hone

Selection of:

Tooling. Fig. 3.21, honing unit selection guide (for internal honing), shows how the shape and length of the hole determine the type of honing tool. Once the type of tool has been decided on, the diameter of the hole determines the individual tool for the job. A honing supply catalog is a guide to the best tool for the workpiece to be honed. Internal honing tools come in 5 categories:

1. Single-stone tools for small holes, about .060 to 1 inch (1.5 to 25mm) in diameter. Longer stones and longer shanks are available for longer or recessed holes.
2. Multiple-stone tools for larger holes, about .625 to 6 inches (16 to 150mm) in diameter. (See Fig. 13.23.) Longer tools are available, with more stones in line.
3. Wide stone tools for honing over keyways and splines. (See Fig. 13.24.) For holes from .250 to about 4 inches (6 to 100mm) in diameter.
4. For large holes, about 2.9 to 12 inches (73 to 305mm) in diameter, it is more practical to use a tool that has more than 1 expanding contact point, such as this 6-point tool. There is practically no length limit of parts to be honed by this type to tool. The parts suitable for honing with all tools listed so far are usually light enough to float on the rigidly mounted rotating tool. But this multi-point tool is for workpieces that are generally so heavy that the part is fixed and the tool floats on 2 universal joints as the honing stones expand in the hole being honed.

BORE DESCRIPTION AND HONING UNIT REQUIREMENTS

Fig. 13.21   Honing unit selection guide

OPEN HOLES WITH NO INTERRUPTIONS

SELECT A HONING UNIT WITH STONE LENGTH 2/3 TO 1-1/2 TIMES BORE LENGTH.

OPEN HOLES WITH KEYWAYS OR SPLINES

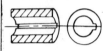

BORES WITH INTERRUPTIONS SUCH AS THESE REQUIRE KEYWAY HONING UNITS. STONE SHOULD BE 2/3 TO 1-1/2 TIMES BORE LENGTH. NOT SUITABLE FOR BLIND HOLE WORK.

OPEN HOLES WITH TANDEM LANDS

TANDEM DISTANCE

TANDEM BORES REQUIRE A STONE LENGTH AT LEAST TWICE THE TANDEM DISTANCE. IF HONING UNITS ARE NOT AVAILABLE WITH SUFFICIENT STONE LENGTH, ALTER THE STONE.

BLIND HOLES

RELIEF

BE SURE SHANK IS LONG ENOUGH TO PERMIT STONE TO REACH BOTTOM IN HOLE.

BOWED HOLES

FOR BOW CORRECTION USE A HONING UNIT WITH A LONG STONE, PREFERABLY 1-1/2 TIMES AS LONG AS THE BORE. BE SURE SHANK IS LONG ENOUGH TO PERMIT CORRECT STROKING.

HOLES WITH COUNTERBORE OR OVERHANG

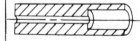

MANDREL SHANK MUST BE LONG ENOUGGH TO REACH THE SURFACE TO BE HONED AND PERMIT OVERSTROKE OF APPROXIMATELY 1/3 STONE LENGTH. IF COUNTERBORE IS ONLY ON ONE END, AND TOLERANCES ARE NOT CRITICAL, PART MAY BE HONED FROM OPPOSITE END ONLY AND HONING UNIT SELECTION MADE AS FOR A PLAIN OPEN HOLE.

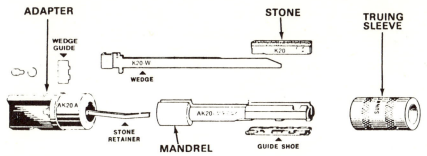

Fig. 13.22   Single stone honing tool

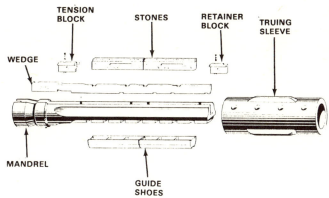

Fig. 13.23   Multiple stone honing tool

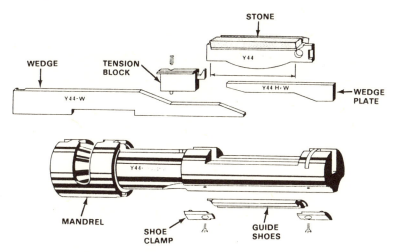

Fig. 13.24   Wide stone honing tool

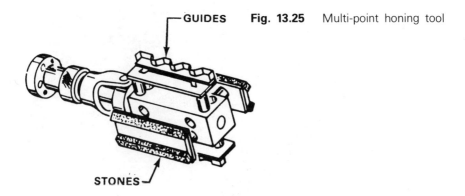

GUIDES   **Fig. 13.25**   Multi-point honing tool

STONES

5. Single-stroke honing tools for holes from .235 to 2 inches (6 to 50mm) in diameter. (See Figs. 13.26, 13.27.) They are very fast and accurate but limited in metal removal. Light workpieces can float; heavy parts will have to be positioned accurately (within .001 in. or 0.30mm) and then locked in place. That is a significant departure from multi-stroke honing procedure.

A single-stroke honing tool consists of an expandable diamond-plated sleeve on a tapered arbor. During setup the plated sleeve is expanded by being pushed up the tapered arbor. The single-stroke honing process is a fast and accurate method of honing a bore to final size. It is fast because the rotating single-stroke honing tool is pushed through the bore only once, unlike conventional honing, where the workpiece is stroked repeatedly over the rotating honing tool. Repeatability of size is very good, because all size adjustments are done only once, during setup.

The operator simply loads and unloads parts. With diamond abrasive plated to that tool, as many as 100,000 parts can be honed with one tool.

Part: hydraulic valve body
Material: cast iron, soft

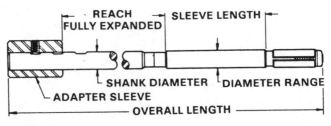

REACH   SLEEVE LENGTH
FULLY EXPANDED

SHANK DIAMETER   DIAMETER RANGE
ADAPTER SLEEVE
OVERALL LENGTH

Fig. 13.26   Single-stroke honing tool

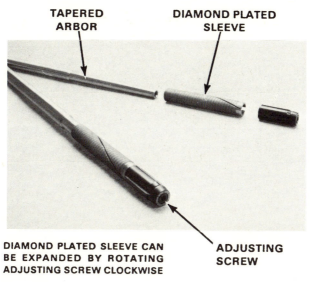

TAPERED ARBOR

DIAMOND PLATED SLEEVE

DIAMOND PLATED SLEEVE CAN BE EXPANDED BY ROTATING ADJUSTING SCREW CLOCKWISE

ADJUSTING SCREW

Fig. 13.27   Single-stroke honing tool

Prior operation: bore
Stock removal: .0025in. (0.064mm)
Honing time: 10 seconds
Accuracy after honing: .000,050in. (0.001mm)

This hydraulic valve body is an example of a workpiece that is well suited for the single-stroke honing process. Cast iron makes short, non-stringy chips. Interrupted or short bores produce a low overall volume of chips and allow the honing oil to wash the chips off the tools during honing. Hard steel, soft steel, bronze, aluminum, and ceramics are also being sized by the single-stroke honing process, but their stringy chips require somewhat reduced stock removal, compared with cast iron.

The volume of chips has to be kept relatively small. That is very important, because, contrary to multi-stroke conventional honing, the single-stroke honing tool cannot take more chips out of the hole than there is room for between the diamond grits. Therefore, in jobs requiring large amounts of stock to be removed and in parts with long, uninterrupted holes multi-stroke conventional honing should be the choice.

Stones.   This sample page from a honing supplies catalog shows that the easy way is to follow the supplier's advice and pick a recommended stone, based on material to be honed, hardness, hole condition, and surface finish desired.

However, when it comes to honing hard steel, the sample page does

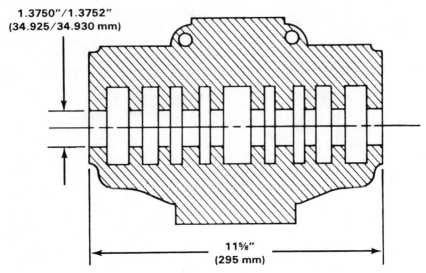

1.3750"/1.3752"
(34.925/34.930 mm)

11⅝"
(295 mm)

Fig. 13.28   Hydraulic valve body

not give a definite recommendation; instead it offers a choice between conventional and superabrasive stones. Another chart suggests using conventional abrasive if only a few parts have to be honed but to use superabrasive for large quantities of workpieces. The reason for that apparent indecision is the price difference between conventional and superabrasives. Borazon grit is manufactured the same way man-made diamonds are made, with tremendously high temperatures and pressures, and it sells for the same price as diamond. Therefore a superabrasive honing stone may sell for 50 times as much as the same size conventional abrasive stone, and the choice is difficult.

However, for hard steel the best choice is usually the superabrasive. But the only definite way to find the best stone is to run a honing test and indicate the results in a comparison chart as shown below:

Two lots of identical M5 hard steel parts were honed. One lot was honed with ordinary abrasive stones and the other with resin bond Borazon. Borazon honed 8 times as fast as ordinary abrasive. While the chart shows the abrasive cost using Borazon was 3 times the cost per part using ordinary abrasive, the saving in labor cost is so great that the total cost per part is only ¼ as much as the total cost with ordinary abrasive.

In soft steel, on the other hand, the primary advantage of Borazon is that of stone life. While cutting rates may be somewhat improved over

NOTE THE CLOSELY PACKED DIAMONDS AND THE LIMITED SPACES FOR CHIPS.

Fig. 13.29   Electron micrograph of plated diamond on single-stroke honing tool (x600)

conventional abrasives, the very slow wear of those metal bond super-abrasive stones causes each successive part to be practically the same size as the one before, and that automatic size control represents the major economic advantage.

Note that metal bond Borazon is best for soft steel and cast iron. Hard steel can also be honed with metal bond, but resin bond may have to be used if metal bond does not cut. That is likely to happen when one is honing long parts with large abrasive contact areas.

Machine Settings

Spindle speed.   Grinding wheels run at up to 7,500 surface ft. per minute (about 2,300 meters per minute). Compared with grinding, honing surface speed is slow, about 200 surface ft. per minute (60 meters per minute) with aluminum oxide or silicon carbide, or 250 sfm (75 m/min) when one is honing with superabrasives (Borazon or diamond).

A convenient way to find the correct spindle speed for a given diameter is

800 divided by the diameter in inches.
Example: 800 divided by 4 inches = 200 rpm; or
20000 divided by the diameter in millimeters.
Example: 20000 divided by 100 mm = 200 rpm.

Table 13—2   Comparison of costs for honing tests

Stone	Time for .001" Stock Removal	Stone Wear Per .001" Stock Removal	Labor Cost Per Part at $.01/sec.	Stone Cost Per Part	Total Cost Per Part	
Borazon™ $114.00 per stone	15 sec.	.000030" =3834 parts per stone	$.15	$.03 per part	$.18	1/7 total cost of Silicon Carbide
Silicon Carbide $2.40 per stone	2 min.	.0005" =249 parts per stone	$1.20	$.01 per part	$1.21	7 times more expensive than Borazon™

Stroke Rate.   A characteristic feature of a honed surface is crosshatch, which makes an excellent oil retention and bearing surface. (See Fig. 13.31.) The crosshatch pattern is generated on the bore surface as the workpiece is stroked back and forth over the rotating honing tool. The faster the stroking rate (spm) in relation to the tool rotation (rpm) the larger the crosshatch angle. Athough exact crosshatch angles can be generated, most blueprints ignore that angle, as it has never been proven to make a difference in the functioning of a honed workpiece. For fastest stock removal, generate the largest possible crosshatch angle. Limitations in the stroke speed capability of the honing machine will make it highly unlikely that the crosshatch angle will be too large.

Stroke length.   The stroke length is not very important if the honing stone is longer than the diameter of the hole being honed. See Figs. 13.22 and 13.23 for examples of such tools. On the other hand, honing tools used for large diameters, such as shown in Fig. 13.24 are *very sensitive* to stroke length. Changing the stroke length by as little as $\frac{1}{8}$ inch (3mm) on each end can make the difference between bellmouth and barrel shape.

On any honing job the abrasive has to work the entire length of the hole plus a little extra on each end, which is called the overstroke.

Overstroke each end by about $\frac{1}{4}$ of the bore length or tool length whichever is shorter.

Feed (pressure).   Any multi-stroke honing operation requires that the stone be expanded against the wall of the hole being honed and a force

HONING STONES FOR BOTH K20 AND J-K20 MANDRELS

> In some cases stones other than the RECOMMENDED STONES may hone faster or last longer. For long or repetitive production runs, it may be economical to choose a stone slightly harder or softer, coarser or finer. As a general rule, hard materials require soft stones, soft materials require hard stones, rough holes require hard stones. See Page 4 for Approximate Surface Finish Chart

RECOMMENDED STONES

OPERATIONS and MATERIALS	Most Commonly Used Stones	Approximate Microinch Surface Finish
DEBURRING in Rough Holes, All Materials	K20-A413	
FAST STOCK REMOVAL in Deburred, Bored, Ground, or Reamed Holes—		
Aluminum	K20-J57	33
Brass, Soft	K20-J63	33
Bronze	K20-J57	55
Carbide	K20-DV57	30
Cast Iron	K20-J57	20
Ceramic	K20-DV57	50
Glass	K20-DV57	95
Steel, Soft	K20-A57	25
Steel, Hardened (try this stone first)	K20-A55	18
Steel, Hardened (if A55 does not cut)	K20-A63	12
Steel, Very Hard Tool (if A63 does not cut)	K20-J63 or K20-NR53	12 or 30
FINE FINISHING in Previously Honed Holes—		
Aluminum	K20-J95	6
Brass, Soft	K20-J83	15
Bronze	K20-J95	12
Carbide	K20-DV07	3
Cast Iron	K20-J95	5
Ceramic	K20-DV07	15
Glass	K20-DV07	15
Steel, Hard	K20-J83	5
Steel, Soft	K20-J95	4

ALL AVAILABLE STONES

ALUMINUM OXIDE STONES (A)—12 per box

STONE HARDNESS	GRIT SIZE							
	80	150	220	280	320	400	500	600
SOFT		K20-A43	K20-A53	K20-A61	K20-A73			
		K20-A45	K20-A55	K20-A63	K20-A75			
			K20-A56	K20-A65				
		K20-A47	K20-A57	K20-A66	K20-A77			
			K20-A58	K20-A67				
		K20-A49	K20-A59	K20-A68	K20-A79			
		K20-A411		K20-A69				
HARD		K20-A413						

SILICON CARBIDE STONES (J)—12 per box

STONE HARDNESS	GRIT SIZE							
	80	150	220	280	320	400	500	600
SOFT	K20-J23	K20-J43	K20-J55	K20-J63		K20-J83	K20-J93	K20-C05*
	K20-J25	K20-J45	K20-J56	K20-J65		K20-J85	K20-J95	
			K20-J57	K20-J66			K20-J97	
	K20-J27	K20-J47	K20-J58	K20-J67		K20-J87		
		K20-J48	K20-J59	K20-J68			K20-J99	
		K20-J49		K20-J69		K20-J89	K20-J911	
HARD				K20-J611				

DIAMOND (D) AND CBN/BORAZON STONES (N)

M - Metal Bond   R - Resin Bond
V - Vitrified Bond—1 per box

	GRIT SIZE								
	70	100	150	220	280	320	400	500	600
			K20-DV47	K20-DV57			K20-DV87		K20-DV07
				K20-DV59					K20-DV09
				K20-NR51			K20-NR83		
				K20-NR53					
	K20-NM15	K20-NM35	K20-NM45	K20-NM55			K20-NM85		
	K20-NM17	K20-NM37	K20-NM47	K20-NM57			K20-NM87		
	K20-NM19	K20-NM39	K20-NM49	K20-NM59			K20-NM89		
				K20-NV59			K20-NV89		

*NOTE: For best results use with Bronze Mandrel.

Fig. 13.30 Honing supplies catalogue, sample page

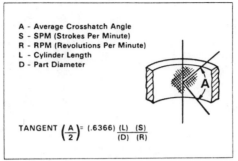

Fig. 13.31   Crosshatch honing angle

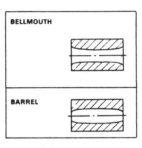

Fig. 13.32   Bellmouth vs. barrel shape bores

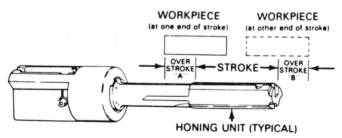

Fig. 13.33   Honing overstroke

applied to make the abrasive cut the material. The stone must then be fed out further to compensate for stone wear and the ever-increasing diameter of the hole. That feed motion can be obtained in several ways by a wedge or a rack-and-pinion arrangement, for example. Once the stone has been fed out against the part wall, it has to exert the correct amount of force to keep cutting. Too little force (or cutting pressure, as it is usually called) will allow abrasive grains to become dull and stop cutting. Too much force will make the stone wear too fast (which is expensive). It will also tend to impair geometric accuracy.

It is very important to find the correct feed force. Use the lowest cutting pressure that gives good cutting action. To determine what "good" is in case of production honing, try different pressure settings and tabulate the results. As Table 13—3 shows, cutting pressure 3 is the one to use in this example:

Table 13—3   Cutting pressure vs. cost

Cutting Press.	Seconds for .004* Stock Removal	Stone Wear Per Part	Stone Cost Per Part	Labor Cost Per Part at 1C/sec	Total Cost Per Part
2	30	.0001	1 cent	30 cents	31 cents
2½	20	.0003	3 cents	20 cents	23 cents
3	10	.0005	5 cents	10 cents	15 cents
3½	5	.0015	15 cents	5 cents	20 cents

Oil.   The oil nozzles should be so arranged that the honing oil enters the workpiece parallel to the honing tool. For small-diameter holes better straightness will result from using 2 nozzles, 1 from each end.

When one is honing long tubes, it is better to tilt the tube slightly and flow oil from the higher end only. That helps to wash the metal chips from the hole. In the case of tube honing, the chips can amount to a considerable volume.

It sometimes is considered desirable to refrigerate the honing oil, but tests have shown that the honing action is actually faster with hot than with cold oil. Another idea was that cold oil would guarantee the exact size of the finished part without one's having to consider the shrinking of the hole diameter when the oil cooled off. That was also unsuccessful, because the input of temperature during rapid honing is vastly greater than the very limited cooling ability of oil.

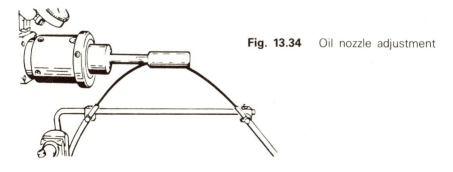

Fig. 13.34   Oil nozzle adjustment

Results of Honing

Surface Finish

Roughness. The finer the grit of the honing stone, the finer the honed finish. The rougher the grit, the lower the cost of honing. If a fine finish is required, rough hone first; then finish hone. To calculate the amount of stock removal required to change from 1 finish to another, subtract the finish desired from that produced by the roughing stone and divide by 10. The answer is the number of .0001 inch (tenths) required (as measured on the diameter) to change from 1 finish to another. For example, if the roughing stone produces 50 microinches and the finishing stone 10, the difference is 40. Divide 40 by 10, and the required stock removal in that case is $\frac{4}{10}$ (.0004 inch). Removing less stock will not produce the desired finish, and removing more is an unnecessary expense.

The metric version of that experience value is similar, except that the difference between rough and fine Ra should be divided by 100. For example, 1.25 micron Ra rough finish minus 0.25 micron Ra fine finish = 1 micron Ra divided by 100 is 0.01 micron stock removal required, as measured on the diameter.

Plateau Honing. A special finish has generated interest in the engine manufacturing and rebuilding market. With that finish the valleys are deep, and the peaks have been removed to form plateaus, giving the name plateau honing or plateau finish. The proponents for the system claim it to be good because:

(1) A cylinder wall has a plateau finish after the engine has been broken in. The peaks of the normally honed surface have been worn off by the piston rings, so you will get a plateaued surface eventually, whether you want it or not. But if you plateau hone originally, then the worn-off peaks of metal are left in the honing machine, instead of in the lubricating oil in the engine.

(2) The valleys act as oil reservoirs and provide lubrication during the critical moments when the engine has just been started, and the oil is not yet at full pressure. A smooth surface without those valleys does not have that ability to retain oil.

A test by a piston ring manufacturer, using a sophisticated radiometeric oil consumption measurement technique, compared 4 thinwall 400 cubic inch Chevrolet engines. Three engines had been honed to produce conventional finishes, but the fourth engine had been honed to have a plateau finish. The engine with the plateau finish consumed $\frac{1}{10}$ the oil that the engines with conventional finishes did and had 80% less cylinder bore wear.

Table 13—4 Approximate surface finish in micro inches (μ^{11})

Material	Stones	Grit Size								
		80	100	150	220	280	320	400	500	600
Hard Steel	Aluminum Oxide/Silicon Carbide	25	—	20	18	12	10	5	3	1
	CBN/Borazon™	—	55	45	30	28	—	20	—	7
Soft Steel	Aluminum Oxide/Silicon Carbide	80	—	35*55	25	20*35	16	7*10	4*8	2
	CBN/Borazon™	—	65	—	—	—	—	25	—	—
Cast Iron	Silicon Carbide	100	—	30*40	20	12	10	6	5	3
	Diamond	—	—	—	80	—	—	50	—	20
Alum, Brass Bronze	Silicon Carbide	170	—	80	55	33	27	15	12	2
Carbide	Diamond	—	—	30	20	—	—	7	—	3
Ceramic	Diamond	—	—	50	40	—	—	20	—	15
Glass	Diamond	—	—	95	70	—	—	30	—	15

*If 2 values are shown: the first number is for small parts, honed on machines with 1 hp or less; the second number is for large parts, honed on machines with 2 or more hp.

Table 13—5 Approximate surface finish in micrometers (μm)RA

Material	Stones	Grit Size								
		80	100	150	220	280	320	400	500	600
Hard Steel	Aluminum Oxide/Silicon Carbide	1.00	—	0.80	0.70	0.50	0.40	0.20	0.10	0.05
	CBN/Borazon™	—	2.20	1.80	1.20	1.10	—	0.80	—	0.30
Soft Steel	Aluminum Oxide/Silicon Carbide	3.20	—	1.40*2.20	1.00	0.80*1.40	0.60	0.30*0.40	0.20*0.30	0.10
	CBN/Borazon™	—	2.60	—	—	—	—	1.00	—	—
Cast Iron	Silicon Carbide	4.00	—	1.20*1.60	0.80	0.50	0.40	0.25	0.20	0.10
	Diamond	—	—	—	3.20	—	—	2.00	—	0.80
Alum, Brass Bronze	Silicon Carbide	6.70	—	3.20	2.20	1.30	1.10	0.60	0.50	0.10
Carbide	Diamond	—	—	1.20	0.80	—	—	0.30	—	0.10
Ceramic	Diamond	—	—	2.00	1.60	—	—	0.80	—	0.60
Glass	Diamond	—	—	3.80	2.80	—	—	1.20	—	0.60

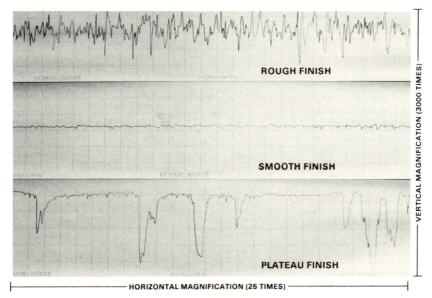

Fig. 13.35   Plateau finish vs. smooth and rough finish

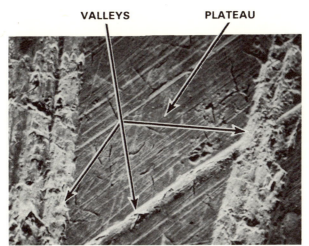

Fig. 13.36   Electron micrograph of a honed plateau finish (x200)

Table 13—6 Stock removal rates

Material	Hardness Rockwell C	Dia. Inch/mm	Length Inch/mm	Stock Removal Inch/mm	Time Sec.
Stellite	50	.129	.375	.001	20
		3.28	9.50	0.025	
AISI 8620 Steel	60	.595	.875	.001	15
		15.11	22.20	0.025	

Material	Hardness Rockwell C	Dia. Inch/mm	Length Inch/mm	Stock Removal Inch/mm	Time Sec.
D2 High Speed Tool Steel	62	.437 11.10	2.250 57.20	.004 0.100	20
Malleable Iron	Soft	1.125 28.58	1.125 28.60	.0013 0.100	35

Material	Hardness Rockwell C	Dia. Inch/mm	Length Inch/mm	Stock Removal Inch/mm	Time Sec.
Ledloy Steel	Soft	1.000 / 25.40	4.000 / 101.9 Blind	.0025 / 0.064	20
303 Stainless Steel	Soft	.437 / 11.10	.750 / 19.10	.003 / 0.076	25

Material	Hardness Rockwell C	Dia. Inch/mm	Length Inch/mm	Stock Removal Inch/mm	Time Sec.
CD-25 Carbide Tungsten	Hard	1.062 / 26.98	1.218 / 31.00	.035 / 0.889	420
6061-T6 Aluminum	Soft	2.375 / 60.33	.537 / 13.6	.007 / 0.178	120

Material	Hardness Rockwell C	Dia. Inch/mm	Length Inch/mm	Stock Removal Inch/mm	Time Sec.
C-1144	Soft	.687	.400	.0025	20
		17.45	10.2	0.064	

Table 13—7 Troubleshooting guide

TROUBLESHOOTING GUIDE

CONDITION TO BE CORRECTED		Step 1	Step 2	Step 3	Step 4	Step 5
STONE NOT CUTTING	STONE GLAZED* Stone surface looks clean but cutting grains are dulled	Sharpen stone with dressing stick	Increase cutting pressure	Increase stroking speed	Use a softer stone	Check oil to be sure you are using Industrial Honing Oil*
	STONE LOADED* Stone surface looks smeared and clogged with chips	Clean stone with dressing stick	Increase stroking speed	Use a softer stone	Use a coarser grit stone	Check oil to be sure you are using Industrial Honing Oil*
SLOW STOCK REMOVAL*		Increase spindle speed	Increase cutting pressure	Check oil to be sure you are using Industrial Honing Oil*	Use a softer stone	Use a coarser grit stone
POOR STONE LIFE*		Decrease cutting pressure	Use faster spindle speed	Use harder stone	Use coarser grit stone	Check oil to be sure you are using Industrial Honing Oil*
BELLMOUTH**		True stone and shoes with truing sleeve. If part is short or unbalanced, shorten stroke length	Use softer stone	If Bore is LONGER than 2/3 stone length:	Shorten STONE only (or row of stones) slightly on both ends	If bellmouth persists, shorten stones still more but do not shorten shoes any further CAUTION: OVER-CORRECTION of bellmouth will lead to barrel condition
				If Bore is SHORTER than 2/3 stone length:	Shorten STONES AND SHOES equally to 1-1/2 times bore length	
BARREL**		True stone and shoes with truing sleeve	Use finer grit stone	Use longer stone or shorten guide shoes on both ends	Use mandrel with longer stone and shoe	CAUTION: OVER-CORRECTION of barrel condition will lead to bellmouth
TAPER IN OPEN HOLE		True stone and shoes with truing sleeve	Change stroke so tight end of the bore is stroked over the stone farther	Reverse work on mandrel more often	If power stroking, make sure spindle and stroker are in alignment	
TAPER IN BLIND HOLE		Shorten stone and shoe to about 3/4 length of bore - shorten stone more if taper persists	True stone and shoes frequently with truing sleeve	If hole has insufficient or no relief at bottom, use hard tip stone	Provide sufficient relief at bottom of hole	Provide adequate oil flow at bottom of hole to wash cuttings out
OUT-OF-ROUND		Make sure honing tool is recommended size for diameter to be honed	Thoroughly true stone and shoes to exact hole diameter	If thinwall part, decrease cutting pressure	If stone stops cutting at decreased pressure, use softer stone	If power stroking, make sure spindle and stroker are in alignment
WAVINESS		Use honing tool with sufficient stone length to bridge waviness (or lands and ports in bore)				
RAINBOW		Stone length should be at least 1-1/2 times the length of the bore for best bow correction.		Use shorter stroke length (less overstroke)	Use softer stone to avoid part or tool flexing	
FINISH TOO ROUGH*		Decrease cutting pressure	Use finer grit stone	Check oil to be sure you are using Industrial Honing Oil*	Thoroughly true shoes to exact hole diameter	For extremely fine finishes in soft or exotic materials, use bronze mandrel or bronze shoes
RANDOM SCRATCHES in bore*		Decrease cutting pressure	Use finer grit stone	Use softer stone	If hard steel mandrel is being used, change to soft steel mandrel. If soft steel mandrel or shoes are being used, change to bronze mandrel or shoes.	Check oil to be sure you are using Industrial Honing Oil*

*Many honing problems, such as poor cutting action, poor stone life, and rough finish, are caused by the wrong honing oil, insufficient honing oil, dirty honing oil, or contaminated honing oil. Use only clean, full-strength Industrial Honing Oil. Make sure that the honing oil is neither diluted nor "cut" with other oils. Keep solvents and cleaning fluids away from the hone.

**This information applies only to honing tools with a stone length that is a multiple of the hole diameter. For other tools see page __, "Stroke Length."

To produce a plateau finish, rough hone to final size with a 70- to 100-grit stone and finish with a 600-grit stone for 15 to 45 seconds. The amount of time used to hone with the 600-grit stone depends on the amount of plateauing desired. The plateauing operation, with the 600-grit stone, removes so little metal that the diameter of the bore does not measurably increase.

Stock Removal Rates

For large holes it is fairly easy to predict the honing time required. Example: Honing a soft tube, 4 inches in diameter, 48 inches long, stock removal .011 in. on the diameter, using a honing machine with 3 hp at the spindle:

(diameter × length × stock removal)/.21 = honing time in minutes
    Example: (4 × 48 × .011)/.21 = 10 minutes.

The metric version of this formula is:

(diameter × length × stock removal)/3400 = honing time in minutes.
    Example: (100 × 1200 × .28)/3400 = 10 minutes.

That time can be shortened by using more horsepower if the honing machine can provide more power, if the honing tool can withstand more power, and if fixturing can be arranged to take more power without distorting the workpiece.

For small diameter parts it is more difficult to predict honing time. The main limiting factor for stock removal speed is the strength of the honing tool. But stock removal required in small parts is usually less than in large-diameter parts, so honing time is surprisingly equal on that variety of parts.

CHAPTER 14

Adaptive Control in Grinding

Robert S. Hahn
Hahn Associates

Editor's Note

One other driving force behind the need for adaptive control in grinding that must be added to Dr. Hahn's position is the industrial move toward higher levels of automation through the use of increasingly sophisticated computer techniques. Grinding, milling, turning, inspection, and logistic systems are being integrated with a drive toward greater throughput. World competition is demanding faster and faster process schedules. High-speed machining techniques are, therefore, being seen more frequently as a requirement in machine specifications. As the speed of process increases, the manual capability to control decreases and the need for adaptive control increases. At some point, the ability to adaptively control becomes one of the limitations to the ability of the manufacturing industry to automate.

R. I. King

The Need for Adaptive Control

The input variables to grinding machines, as described in chapters 1 and 2, are typically the rough and finish feedrates, amount of finish stock, sparkout time, dressing interval, and dressing parameters. As shown earlier, they affect the 2 important inputs to the grinding process; namely, the normal force at the wheelwork contact, F_n, and the topography of the wheel face or wheel sharpness, which is indicated by the

337

WRP (see Chapter 1). Those 2 process variables govern all of the output variables; i.e., stock-removal rate, Z'_w, surface finish, R_a, surface integrity, and, ultimately, cost per piece.

In addition to the 2 grinding process input variables, F_n, and WRP, random stock variations, random stock runout variations, variations in microstructure and workpiece hardness are also imposed on the grinding operation with the expectation that a high-quality output consisting of uniform size, taper, roundness, surface finish, and surface integrity will be achieved in a short cycle time.

Many grinding systems are able to absorb those random inputs and produce satisfactory output quality. However, as size, taper, roundness, surface finish, and surface integrity tolerances become more stringent and skilled operators become less available, adaptive controls are necessary to produce high-quality output at low cost.

In conventional grinding systems, the 2 important grinding process variables, F_n and WRP, are usually not well controlled. Generally, the sharpness of the grinding wheel, as measured by WRP, changes during the grinding process. As the wheel dulls, the real area of contact increases (see Chapters 1 and 2), causing the WRP to fall. That, in turn, causes the induced force, F_n, to rise (see Eq. (1.15)). The increasing real area of contact, coupled with the increased normal force, soon exceeds the threshold of thermal damage, and poor surface integrity may result. To ensure good surface integrity, adaptive controls are needed to monitor "on line," the wheel sharpness, to achieve self-sharpening conditions, to limit the increase of interface normal force, and/or to initiate a dressing operation.

In conventional grinding systems that are not highly rigid and are required to produce close size and taper tolerances in a short cycle time, variable deflections of the system at the termination point of the cycle cause size and taper variations. Where in-process gaging can be used, the size error at the gage location can be eliminated, but taper errors or size errors at other locations are not eliminated. On fast-grinding cycles with random input stock variation, the induced normal force may not reach the steady-state value before sparkout begins. In that case size and taper errors can be caused by random initial stock variations unless a long sparkout time is provided. Therefore, adaptive controls are needed to absorb stock variations on low or moderately rigid machines operating on fast cycles.

In grinding operations where it is desired to hold close size tolerances on compliant systems without in-process gaging, variations in the threshold force (see Chapter 1) at the termination point of the grind cycle also cause size errors. Those threshold forces (force below which stock removal ceases) are often significant in operations where the difference of curvature between wheel and work is small (internal grind with large D_e, see Chapters 1 and 2). As the wheel dulls and the equivalent diameter, D_e, varies or both, those forces vary and cause size and taper errors. Again, adaptive controls are needed to compensate for variations in threshold force.

Many precision grinding operations require a high degree of roundness, concentricity or both with other surfaces. The runout or eccentricity of the initial stock may affect the roundness of the finish ground part on fast production cycles. There is often a race between rounding up the part and consuming the available stock. Consequently, the initial stock-runout ratio is an important random input variable imposed on the grinding operation. On moderate- or low-rigidity systems, the feedrate may have to be slowed in order to provide sufficient time for achieving the desired roundness. Adaptive controls are needed to enhance the rounding-up process.

In addition to the ability to adapt to the various random input variables, it is desirable to optimize the grinding machine input parameters to produce parts of given quality level at minimum cost or maximum production rate. Two types of adaptive control have been recognized: (1) where constraints are imposed to limit the excursion of certain variables called ACC control[1], and (2) where some object function is to be maximized or minimized called ACO control. In grinding, it is usually desired to minimize the grinding cost function or maximize the production rate function. In current practice the set of machine input parameters that optimize the process is not easily found. Chapter 1 describes a software package that aids in determining the optimum parameters.

Sensors in Grinding

A variety of sensors have been suggested for grinding operations. In-process gaging is, of course, the most widely used type of sensor for measuring the current size of workpieces. Those gages can produce signals corresponding to the instantaneous size and also signals corresponding to the rate of stock removal from the workpiece $\bar{v}_w$. In cylindrical plunge-grinding operations, the rate of wheelwear, $\bar{v}_s$, can be obtained by subtracting $\bar{v}_w$ from the plunge feedrate, $\bar{v}_f$, according to Eq. (1.1) after the steady elastic deformation state has been reached.

Wheelhead motor-drive torque sensors are also available. Placed in the supply lines to the drive motor, they indicate input power. That signal must be converted to torque at the grinding wheel by taking into account the motor, drive train, and wheelhead bearing efficiency, and the rotational wheelspeed. The tangential force, F_t, can then be found by dividing the torque by the wheel radius (see Eq. (1.21)). Finally, the normal force, F_n, can be found approximately by dividing F_t by an average "friction coefficient" (see Eq. (1.22)). However, the "friction coefficient" is a variable depending upon wheel sharpness and, generally, is not precisely known during production grinding operations.

In internal grinding where in-process gaging is inconvenient, Toenshoff[2] has investigated ultrasonic wall thickness sensors. The ultrasonic transmitter and receiver head is acoustically coupled to the rotating O.D. of the workpiece by a film of liquid coolant. The time delay between the transmitted pulse and the reflected pulse is an indication of the wall thickness. Accuracies of ± lum were obtained. Another sensor is described[3] for detecting when the grinding wheel contacts the workpiece. It allows a rapid feed to find the work surface before shifting to the normal grinding feedrate. That sensor detects the noise generated by the impact of the wheel with the workpiece. Studies have also shown that the phase angle between voltage and current in the wheelhead drive motor is particularly sensitive to initial power consumption at the wheelwork interface[2] and can be used to shift from the approach feedrate to grinding feedrate. The change in phase angle has a time constant on the order of 10ms.

Toenshoff[2] also describes a wheelhead equipped with piezoelectric force sensors to measure the normal grinding force as well as the tangential component. The motorized wheelhead spindle cartridge is mounted in a special sleeve adapted to accept the spindle cartridge at 4 points of support, each support point being a piezoelectric force sensor. Another method of measuring the normal force on internal grinders, where the workpiece is chucked in the workhead, consists of mounting a pair of piezoelectric force sensors just behind the chuck and arranged to measure the instantaneous force components in the plane of rotation. Those components are electronically resolved to give the normal grinding force.

Another method of measuring the normal force in grinding is proposed by Hahn[4]. In that method a pair of noncontacting eddy current displacement sensors are installed within the front-end cap of the wheelhead just in front of the forward wheelhead spindle bearing and arranged to detect microdeflections of the rotating wheelhead spindle. Those noncontacting sensors impinge directly on the rotating spindle with a small clearance and operate in a thermally balanced, push-pull mode.

Another method of measuring the normal grinding force uses strain gages installed in a groove in the outer race of the ball bearings supporting the spindle. Those special bearings were developed in Germany and are commercially available.

Spindles mounted in hydrostatic bearings provide the opportunity to measure the differential pocket pressure between opposed pockets to obtain the radial force. The recently developed magnetic bearing developed in France also provides a similar opportunity.

In addition to normal and tangential force sensors, some developments in sensing surface finish have been presented. Chouinard[5] describes noncontact surface quality inspection methods using electro-optical and laser scanning for assessing surface finish and surface flaws. Kaliszer et al.[6] propose a method of measuring surface finish during the grinding process. They press a cylindrical drum against the rotating workpiece, which assumes a position determined by the highest asperities in the region of contact. The drum also carries a set of sping-loaded styli, which protrude from the drum surface and contact the work surface as the drum rolls against the work. By recording the depth displacements of these styli, an indication of the current roughness can be ascertained. Another device proposed by Nowicki[7] presses a flat block against an element of a cylindrical workpiece. Again, the block is located on the highest asperities along the work surface. In this case the block contains a series of hydraulically loaded plungers equipped with styli that seat themselves at a set of points along the surface. The average position of those plungers relative to the flat block reference surface is a measure of the roughness. The average position is indicated by the volume of oil required to seat all of the plungers.

A sensor for measuring wheelwear during the grinding process was described by He Xiu-Shou[9]. The measuring device consists of a plunger having on one end an air nozzle that locates itself on an air film against the moving surface of the grinding wheel. As the wheel wears, the air nozzle follows the surface of the wheel.

Vibration pickups and microphones have been applied in various laboratories to determine the onset of grinding chatter[8]. Wheel spindle vibration is preferred over tail-stock pickups, since it is independent of the workpiece and workpiece tooling.

Adaptive Control Strategies

The incoming stock to production grinders often varies in a random manner. If the grinding system is relatively stiff and long sparkout times are acceptable (over 3 time constants, see Eq. (1.26)), adaptive control is not required. However, if fast cycle time is desired on nonrigid systems, the ability to adapt to stock variations can improve quality. The Heald controlled force internal grinder[9,10] absorbs stock variations during the roughing phase and provides a sparkout or finish phase that always commences with the same normal force regardless of the amount of stock. It eliminates size errors caused by stock variations on fast cycles.

An adaptive control for external cylindrical grinding of air-craft engine rotors has been described by He Xiu-Shou[8]. That system utilized a normal force sensor and computer control of the force during rough grinding. The wheelwear is monitored by a pneumatic wheelwear sensor described above, and the vibration or wheel life is monitored with a vibration pickup. The closed-loop computer control minimizes an empirical function relating the normal force to grinding time plus dressing time to obtain the optimum normal force.

Malkin et al.[11,12,13] has developed a strategy for optimizing both the grinding and dressing parameters for maximum removal rate subject to thermal damage and surface finish constraints. That strategy is the basis for both on-line and off-line optimization systems for plunge-grinding operations on steels. Neither system requires a data base, as direct measurements of power and surface finish are used for identifying and updating the unknown parameters in the grinding model.

A computer program written in BASIC was developed for off-line interactive optimization, using a desktop microcomputer[12], and the same program can also be used to evaluate the efficiency of existing operations. The program user inputs the grinding and dressing parameters, the maximum allowable surface finish, and the measured power and surface finish. The response on the computer screen displays the estimated optimal grinding and dressing parameters, suggested new trial conditions, and the present grinding efficiency, which is the existing removal rate divided by the estimated optimal removal rate. For a number of industrial grinding operations evaluated in that way, the estimated efficiency is óften found to be about 50%. That computer program has been practically applied to plunge-grinding operations in the automobile and rolling-element-bearing industries.

An adaptive control optimization (ACO) grinding system using the same optimization strategy has been developed. It consists of a cylindrical grinder interfaced to a computer[11]. The initial prototype system actually operates in a mixture of on-line and off-line modes. The grinding power measured in-process is fed to the computer, and the computer controls the grinding parameters. The surface finish is measured off-line and is manually input to the computer. However, the availability of a reliable in-process surface finish transducer would eliminate that task, and the surface finish could be directly fed to the computer, thereby enabling complete closed-loop optimization. A more advanced ACO system is being developed that couples the same optimization strategy with an accelerated sparkout strategy to recover the elastic deflection of the grinding system as quickly as possible at the end of the grinding cycle[13].

Hahn and Graham[4] describe the application of an adaptive-control based upon the normal force for improving the accuracy, quality, and production rate of nonrigid production grinding systems. The control uses normal force sensors built into the wheelhead to give a very sensitive measure of normal force. It allows the system to adapt to variations in stock allowance, stock runout, workpiece hardness variations, and wheel sharpness variations to prevent size and taper errors in compliant systems. That is accomplished by compensating for variable deflections in the system caused by threshold force variations (see Chapters 1 and 2). The wheel sharpness is also monitored during the grinding process, and the buildup of normal force caused by dulling of the wheel is prevented to provide control of surface integrity. On low-rigidity systems, the control has the ability to induce artificial rigidity into the system to obtain a faster roundup of eccentric stock[16] (see also Chapter 7). Size and taper errors often occur in internal grinding operations as the wheel size changes from "new-wheel" size to "discard" size. The control causes the cycle parameters to gradually change as the wheel size changes to give more consistent results over the life of the wheel and to produce more parts per wheel.

Smith[14] offers an "energy adaptive" system for maintaining sharp grinding wheels and avoiding thermal damage on external cylindrical grinders. In that system a probe is fed radially into the workpiece to generate a stock-removal rate signal that the closed-loop control maintains constant by feeding the wheel into the workpiece on the opposite side from the probe. As the wheel dulls, the induced forces rise (see Eq. (1.15)), and increased wheelhead power is picked up by the power sensor which, in turn, reduces the wheelspeed. As the wheelspeed is diminished, the "work-removal parameter" (WRP) falls, causing an even greater increase in the induced normal force, which brings the wheel into a rapid wear and self-sharpening region. As the wheel restores its sharpness, the WRP rises, the induced forces fall, the power falls, and the controller increases the wheelspeed to its original value. In that way, the system provides fast stock-removal rates without thermal damage at the expense of increased wheelwear rate and reduced wheel life.

An adaptive control for conveyor-type, coated-abrasive belt grinders is described by Shibata[15]. On conventional belt grinders, the thickness dimension of the blocklike parts being ground varies as a result of variations in the initial stock allowance, variations in sharpness of the belt, and progressive belt wear. In order to improve the sizing accuracy and to extend the belt life, an adaptive control for changing the belt speed as well as the table speed was developed. Several belt and table speed functions were evaluated starting from, initially, a high-belt speed and low-table speed and terminating at a low-belt speed and high-table speed.

Conclusion

Improvements in grinding output quality can be achieved by adaptive control, where random input stock, runout, workpiece hardness, microstructure and wheel sharpness variations are imposed on the grinding system. Adaptive controls also are necessary where highly skilled, diligent operators are not available to continuously monitor the grinding process.

References

1.   E. Salje, H. Mushardt, E. Scherf, - "Optimization of Short Grinding Cycles," C.I.R.P. Annals, Vol. 29/2/2980, pp. 477–95; Hallwag Ltd., Berne

2.   H. K. Toenshoff, G. Rohr, P. G. Althaus, "Process Control in Internal Grinding," C.I.R.P. Annals, Vol. 29/1/1980, pp. 207–11; Hallwag Ltd., Berne.

3.   Anon, Georg Muller, Kugellagerfabrik KG, Nurnberg.

4.   R. S. Hahn, G. Graham, "An Application of Force-Adaptive Grinding" SME Paper MR84-530, Soc. Mfg. Eng's, Dearborn, Mich. 48121.

5.   E. Chouinard, "Current State of the Art in Noncontact Inspection of Ground Surfaces," SME Paper MR84-537, Soc. Mfg. Eng's, Dearborn, Mich. 48121.

6.   H. Kaliszer, H. Fletcher, J. R. Adams, "In-Process Measurement of Surface Roughness during Plunge Grinding," Dept. of Mech. Eng., Univ. of Birmingham, 1985.

7.   B. Nowicki, "Surface Roughness Measurement with New Contact Method," Proc. 25th Int. Machine Tool Design and Research Conf., Univ. of Birmingham, 1985.

8.   He Xiu-Shou, "The Research of a Practical Adaptive Control System for External Cylindrical Grinding Process." Proc. 25th Int. Machine Tool Design and Research Conf., Univ. of Birmingham, pp. 169–75, Macmillan, 1985.

9.   R. S. Hahn, "Controlled-Force Grinding—A New Technique for Precision Internal Grinding," J. Eng. for Industry Trans, ASME Series, B, Vol. 86, pp. 287–93; 1964.

10.   R. S. Hahn, "Some Characteristics of Controlled-Force Grinding," Proc. 6th Int. Machine Tool Design and Research Conf., Manchester, 1965, Pergamon Press.

11.   G. Amitay, S. Malkin, Y. Koren, "Adaptive Control Optimization of Grinding," J. Eng. for Industry, Trans. ASME, Feb. 1981, pp. 103–08.

12. S. Malkin, Y. Koren, "Off-Line Grinding Optimization with a Microcomputer," *Annals of C.I.R.P.,* Vol. 29/1/1980, pp. 213–16.

13. S. Malkin, "Grinding Cycle Optimization," *Annals of C.I.R.P.,* Vol. 30/1/1981.

14. R. L. Smith, "Energy Adaptive CNC of Precision Grinding," SME Paper No. MR79–385.

15. J. Shibata, K Chao Kun, I. Inasaki, S. Yonetsu, "Adaptive Control for Conveyor-Type Belt Grinders," *Annals of C.I.R.P.,* Vol. 29/1/1980, pp. 217–20.

16. R. S. Hahn, "An Investigation of a Force-Adaptive, Creep-Feed Control for Improving the Rounding Capability of Flexible Grinding Systems," Proc. NSF, 12th Conf. on Production Research and Technology, Madison, Wis., SME, Dearborn, Mich.

17. Inoue, H., Tamakohri, K., Suto, T., Noguchi, H., Waida, T., Sata, T.: "An Adaptive Control System of Grinding Process. Proc. Int. Conf. on Production Engineering, Tokyo, 1974.

18. Werner, G., *Konzept und technologische Grundlagen zur adaptiven Prozessoptimierung des Aussenrundschleifens.* Habilitations schrift, TH Aachen, 1976.

19. Salje, E., "Stategien zur Prozessoptimierung beim Aussenrund-Einstechschleifen. Industrie-Anzeiger 95, Nr. 51, 1973.

20. Shaw, M.C., "Cost Reduction in Stock Removal Grinding, *Annals of C.I.R.P.,* Vol. 24/2/1975.

21. Bierlich, R., Technologische Voraussetzungen zum Aufbau eines adaptiven Regelungssystems beim Aussenrund-Einstechschleifen. Dissertation TH Aachen, 1976.

22. Salje, E., Grinding Process, considered as Feedback Control System, *Annals of C.I.R.P.* Vol. 27/1/1978.

23. Salje, E., Roughness Measuring Device for Controlling Grinding Processes, *Annals of C.I.R.P.,* Vol. 28/1/1978.

25. Salje, E., Mushardt, H., Scherf. E., "Optimerregelung fur das Aussenrundeinstechschleifen," VDI-Zeitschrift 122 1980, Nr. 6.

26. Mikhelkevich, V.N., et al.: "Automatic Control of Transverse Feed During Internal Grinding," Machines & Tooling. Vol. XLV, No. 8.

27. Wada, R., "Adaptive Steuerung einer Rundschleifmaschine, Werkstatt und Betrieb 104 (1971), 6.

28. Dutschke, W., Kiesling, W.D., Rau, N., "Oberflachensensor zur Rauheitsmessung beim Aussenrundeinstechschleifen. wt-Z. ind.Fertig. 65 (1975), Nr. 11.

29. Trumpold, H., Mack, R., Messsystem fur die adaptive Regelung der Formabweichung vom Kreis beim Aussenrundeinstechschleifen, *Annals of C.I.R.P.,* Vol. 24/1975.

30. Suto, T., u.a., A Newly Developed In-Process Sensor for Detecting Active Grain Wear in Grinding Operations, *Annals of C.I.R.P. Vol. 25/1/1976.*

31. Storm, T., "Adaptive Control in Cut-Off Grinding," Proc. M.T.D.R. Conf. 1970.

32. Trmal, G., Kaliszer, H., "Analysis of Methods used in the Optimization of Grinding Condition, Proc. NAMRC, 1975.

33.   Konig, W., Bierlich, R., Kleensaug, R., "Development of an Adaptive Control System in Cylindrical Plunge Grinding SME-Paper, Marz 1976.

34.   Cubukov, Kon'sin, "Adaptive Steuerung von Rundschleifmaschinen mit Minicomputern zur Bearbeitung Abgesetzter Wellen," Uberstetzt aus Stanki i instrument (1978), 9.

35.   Ratmirow, V.A., et al.: "Adaptive Control of a Cylindrical Grinding Machine, Machine & Tooling, Vol. 48, No. 8.

36.   Novikov, V.Yu., Bryatova, L.I, "Investigation and Development of an Adaptive Control System for Grinding, *Russian Engineering Journal*, Vol. 57, No. 2, 1977.

37.   Lur'e, G.B., Gichan, V.V., "Adaptive Control System for Plunge Cylindrical Grinding." *Stanki i instrument*, Vol. 45, Issue 7, 1974.

CHAPTER 15

Trouble Shooting Grinding Problems

Robert S. Hahn
Hahn Associates

Grinding problems can usually be classified under one or more of the following headings:

1. Surface Finish
2. Surface Integrity
3. Size Errors
4. Taper, Roundness, and Form Errors
5. Chatter and Vibration

They are discussed below.

Surface Finish/Surface Profile Problems

The surface finish imparted to a ground workpiece depends upon 7 factors; namely: (1) the interface normal force between wheel and work during the last 4 or 5 work revolutions before the termination point in the grind cycle, (2) the condition of the wheel surface, including the grit size, the effective grit spacing, and the wear flats on the grits (degree of wheel dullness), (3) the wheelwork conformity as measured by D_e (see Chapter 1), (4) the sparkout time or number of repetitive overpasses. (5) the cleanliness and type of coolant and its ability to prevent wheel loading and microchip reweld to the work surface or loose grain scratches, (6) the smoothness of the wear flat on the abrasive grain, and (7) the uniformity of the local hardness and structure of the grinding wheel from point to point around its periphery.

The surface finish is strongly dependent upon the normal force (or normal stress) between wheel and work (Item 1). The instantaneous normal force consists of an alternating component superimposed on a steady component. Vibration from the wheel spindle bearings, the drive system, unbalance, self-excited wheel regenerative chatter, and non-uniform wheel hardness around the wheel's circumference contribute to the alternating, instantaneous normal force which, in turn, contributes to the surface roughness.

The steady component of normal force for a given finish feedrate can be reduced by increasing wheel surface speed which, in turn, increases the WRP and causes a drop in induced force (see Eq. (1.15)), thereby improving the surface finish. Increasing the sparkout time also improves surface finish.

The dressing operation and the condition of the dressing device generally affect Item 2, the condition of the wheel surface. Where surface finish problems occur, close attention should be given to the dressing operation. For single-point diamond dressing, the depth of dress and the dressing lead are very important and should be controlled. The sharpness of the dressing tool also matters. Good dressing practice requires that the wheel be dressed in 1 direction only to produce, in effect, a fine "thread" on the surface of the wheel with a pitch on the order of .001 in. If the wheel is dressed on an out stroke followed by an in stroke, 2 fine "threads" are generated of opposite sense, and 2 areas result on the wheel where those threads cross. Those areas produce a different surface finish on the workpiece and result in a nonuniform surface finish pattern.

For rotary diamond dressing tools, vibration during the dressing process can produce a rough wheel surface which, in turn, produces a rough work surface. Vibration between wheel and dresser may be self-excited, similar in nature to lathe chatter in the turning process, or of the forced vibration type. On high-speed spindles where there is some flexibility, the wheel may run out because of unbalance or eccentric drive sheaves. As the wheel is dressed, it will run true only at the position of the diamond dresser. If the work is located at some other angular position around the wheel, a chatter or surface finish pattern may result.

The wheelhead spindle bearings also influence the degree of cylindricity of the dressed wheel. In some wheelheads, depending upon the bearing design, the axis of rotation may precess or wander about the geometric axis. That causes the dressed wheel to resemble, in effect, a stack of washers mounted on a bolt, each washer being slightly misaligned with respect to its neighbors. The surface finish produced on the workpiece by such a wheel under plunge-grinding conditions is significantly poorer than under reciprocating conditions. The ratio of those 2 surface finishes is an indication of the quality of the dressing operation.

The surface finish and the surface profile (peak-to-valley depth over considerable extent) usually deteriorate from the "as dressed" condition as the grinding wheel wears. That deterioration is caused by local variations in wheelwear rate around the circumference of the wheel and is a direct result of local wheel hardness and structure variations in the wheel. The rate of deterioration of surface finish/profile under a given normal force intensity or normal stress is an index of the uniformity and quality of the grinding wheel. The effect of local wheel hardness variations can be observed by taking a series of closely spaced Proficorder or Talysurf charts across a plunge-ground surface and comparing the profile of various longitudinal sections.[7]

Another variable that affects the surface finish is the "Equivalent Diameter," described in Chapter 1. As grinding wheels wear and become smaller, the D_e drops, and the surface finish becomes poor.

Surface Integrity

Thermal damage or grinding burn often occur in grinding operations. As described in Chapter 2, there are 4 grinding variables involved in driving the surface temperature over the critical value which cause the onset of thermal damage. They are (1) the wheel sharpness (measured by WRP), (2) the induced normal force intensity/stress, (3) the length of the wheelwork contact zone or "footprint" governed by the local equivalent diameter, D_e, and (4) the effective "footprint" speed over the workpiece. Of course, properly directed high-pressure coolant into the wheelwork zone plays a vital role in removing heat and is extremely important. Although water-base fluids have a higher specific heat than do oils, they lack the lubricity, the much higher flash point, and the ability to remove heat above the boiling point of water. As a result, grinding oils often inhibit thermal damage more effectively than do water-base fluids.

In order to avoid thermal damage, monitoring the wheel sharpness and limiting the buildup of induced force caused by wheel dulling may be necessary, as outlined in Chapter 14 "Adaptive Control in Grinding." Increasing the dress lead and depth of dress for single-point diamond dressing increases the sharpness of the wheel and reduces the induced normal force/stress. Reducing wheel diameter, or the D_e, helps to prevent thermal damage. The use of open structure wheels and the use of cubic boron nitride (CBN) wheels also result in cooler grinding action.

Holding Close Size Tolerances

Piece-to-piece size errors may be caused by nonrepetitive movements of the machine, or they may be caused by variations in grinding process variables such as: stock variations, stock runout, wheel sharpness variations, and wheelwear variations. When one is using fast production feedrate cycles where stock variations occur, frequently steady-state conditions in the rough grind are not attained, resulting in force and deflection variations at the end of roughing. Those deflection variations are reduced during sparkout. If the sparkout time is short (less than 3 time constants), size errors result. Variations in wheel sharpness also produce force and deflection variations which, again, result in size errors.

Variations in threshold force likewise produce size errors even after long sparkouts.

One way of eliminating those size errors attributable to variations in deflection is to use computer, force-adaptive grinding where a microcomputer interrogates a load cell in the wheelhead and compensates for deflection variations to provide fast cycles of high precision. (See Chapters 7 and 14.)

Taper, Roundness, Form Errors

In cylindrical reciprocating grinding, the wheelhead axis, the workpiece axis, and the direction of reciprocation are nominally parallel to each other. If the wheel spindle and workpiece are essentially rigid and the wheel is dressed to be a true cylinder, a straight cylindrical workpiece will be generated. However, if the wheel support and/or workpiece are not rigid, lateral and angular deflections occur under the grinding force, causing the cutting surface of the wheel or the workpiece axis to undergo slight angular deflections resulting in the generation of a taper during the rough grinding part of the cycle. The cutting surface of the wheel at that time is generally not exactly parallel to the reciprocation direction of the table slideway. That causes the normal interface force to pulsate and synchronize with the table stroke, and if that force/stress approaches the wheel breakdown force/stress, local wheelwear may take place, destroying the precise cylindricity of the "as dressed" wheel. As the machine is allowed to spark out, the grinding force decays. If the threshold force is zero, the angular deflections approach zero and no taper results except for that caused by the excessive local wheelwear. On the other hand, where significant threshold forces exist, residual angular deflections cause taper errors.

In order to reduce taper errors, increase the angular stiffness of the system or reduce the threshold forces. Aggressively dressing the wheel to increase its sharpness helps to reduce the threshold forces. Reducing the wheel diameter or D_e also tends to reduce the threshold forces. For some grinding systems, the taper is automatically eliminated with the aid of normal force sensors interfaced to a computer.[2]

Roundness errors sometimes occur in workpieces. A common and obvious cause on chucking machines is due to clamping forces in the chuck. The work is ground round but springs out of round when released from the chuck. That can sometimes be alleviated by clamping the workpieces axially instead of radially.

On workpieces containing residual stress, the roundness of a ground surface may be destroyed as layers of stressed material are ground away. Avoidance of roundness errors may require rough-and-finish operations with an intervening stress-relieving treatment.

Roundness errors may also be generated as a result of asymmetric rigidity in the rotating workpiece system. For example, if the center hole in the workpiece on a center-type grinder is elliptical, the work will have a higher stiffness in 1 direction than at 90° to that direction. As the grinding force scans the rigidity of the rotating work system, it detects 2 highs and 2 lows on every revolution. As a result, the work is ground to an elliptical shape during the rough grind. Again, as the machine sparks out, the roundness is improved until the force drops to the threshold value. Thereafter, there is no further improvement. It is better to support the work on 3 contacts spaced 120° apart to obtain an axisymmetric rigidity in the rotating system.

On fast-grinding cycles where a long sparkout time is not permitted, the wheel depth-of-cut h at the moment of retraction may cause a roundness error (see Eq. (1.8)). Increasing the workspeed will reduce the wheel depth of cut and the roundness error.

On fast-grinding cycles (sparkout time less than 3 time constants), there is often a race between rounding up the initial runout and removing the stock to size. If the part reaches size before it has been completely rounded up, roundness errors occur. The roundup rate depends upon the cutting stiffness K_c and the system stiffness K_s. Sharpening the grinding wheel or reducing its width, reducing D_e tend to reduce K_c and improve the rate of roundup. An adaptive control has also been developed for enhancing the rate of roundup.[1]

On slow-grinding cycles where the sparkout time is greater than 3 time constants (see Eq. (1.26)), roundness errors from the initial runout may still occur when threshold forces exist. The condition for obtaining roundness is that the steady state must be reached on the low spot before sparking out.[2] In that way the roundness is achieved in the cutting region (Fig. 1.05, Chapter 1), and then sparkout drops the forces into the plowing and rubbing regions once the part is round.

Finally, roundness errors may be caused by unbalance in the work-rotating system or by inaccuracies in the workhead bearings.

Form errors or deviations from cylindricity or flatness are sometimes caused by local heating of the workpiece, causing the work surface in the footprint zone to expand. That creates higher normal force or stress, which further increases the expansion and creates a thermally unstable condition. A "relaxation oscillation" results, where the spark stream increases to a maximum and then suddenly collapses as the work loses contact with the wheel on the cooling cycle. That phenomenon generally occurs at low workspeeds, where considerable heat is entering the workpiece. Applying the coolant under high pressure and ensuring that it penetrates to the wheelwork interface tends to prevent those thermal relaxation oscillations. Increasing the workspeed while maintaining the interface force constant also tends to eliminate the effect. Note that increasing the workspeed in creep-feed operations does not maintain the force constant and may not eliminate the effect.

Form errors may also be caused by differential wheelwear rates over the wheel's cutting face. As different areas on the wheel are subjected to different normal-stress levels, differential wear takes place, destroying the form on the wheel. It is to be noted that the wheelwear rate varies between the 2.5 and third power of the normal stress.[3] Increasing the wheelspeed and wheel hardness and reducing the feedrate tend to reduce wheelwear rates.

Chatter and Vibration

Trouble shooting chatter and vibration problems on the shop floor can sometimes be accomplished without the use of highly technical and sophisticated equipment. The first step is to use the chatter pattern on the workpiece to determine the frequency of the vibration. The vibration frequency equals the work surface speed divided by the chatter wavelength, both easily measured quantities.

Grinding chatter patterns can be interpreted and classified into six types. If the frequency corresponds to the wheelspeed, the chatter can be classified into 1 of 3 types:

(1) Wheel or spindle unbalance—a straight-line pattern repeating at wheelspeed frequency;

(2) Geometrical runout of wheel surface at the point of grinding—a straight-line pattern repeating at wheelspeed frequency; and,

(3) Wheel mottle patterns—a nonstraight-line random pattern repeating at wheelspeed frequency.

If the frequency does not correspond to the wheelspeed, it can be classified again into 1 of 3 types:

(4) General forced vibration caused by pulleys, belts, drive motors, hydraulic pumps, etc. In that case the frequency will change in proportion to the speed of these elements;

(5) Wheel regenerative type, in which the straight-line chatter frequency is essentially independent of workspeed, drive motor speed, etc., but corresponds, approximately, to a natural frequency of the wheelhead, workhead, or machine structure. In that type the wheel gradually wears into a multilobed cylinder; and,

(6) Work regenerative type, again, where the straight-line frequency corresponds to a natural frequency but where the workpiece develops a wavy surface similar to a "corduroy road." It tends to occur at high workspeeds.

In the above cases a straight-line chatter occurs in the workpiece under plunge-grinding conditions. If a spiral chatter pattern occurs, vibration between the dresser diamond and the wheel during the dressing process is probably the cause.

Once the type of chatter has been determined, steps can be taken to eliminate it. Since types 1, 2, and 4 are generally well understood, only types 3, 5, and 6 will be discussed below.

Wheel Regenerative Chatter

Wheel regenerative chatter is a type of self-excited vibration and is to be distinguished from forced vibrations. In a self-excited vibration, the periodic driving force is created and controlled by the vibratory motion itself whereas in a forced vibration the driving force is independent of the vibratory motions. During a grinding cycle, if the grinding wheel is given a small vibratory disturbance, the interface force between wheel and work will fluctuate. That causes a local fluctuation of the instantaneous amount of wheelwear. After the wheel has made 1 revolution, this local minute wavy-wheel surface produces a transient force variation which, in turn, may cause another fluctuation in wheelwear. If the system is stable, those disturbances die out. If the system is unstable, they build up. (Snoeys and Brown and others[4,5]) have investigated the stability of grinding. (See Chapter 6.)

Since most practical grinding operations lie in the unstable region and since satisfactory grinding can be accomplished as long as the vibration amplitude is less than a certain value, the concept of a "chatter-free grind time" has been proposed by Lindsay.[6] The objective in production grinding operations is to select grinding conditions so that the CFGT is sufficiently long to accommodate the cycle time. Redressing the grinding wheel, of course, reconditions the wheel, allowing it to start another CFTG.

There are 8 rules for eliminating wheel regenerative chatter:

(1) Dress the wheel more frequently.
(2) Reduce feedrate or force intensity.
(3) Increase D_e; i.e., use a larger wheel if one is internally grinding.
(4) Increase the stiffness of the wheel support and/or work support.
(5) Reduce the width of cut or wheel face.
(6) Reduce the wear rate of the wheel by using a high-performance grinding fluid.
(7) Reduce the workspeed.
(8) If the number of lobes on the wheel are less than 10, adjust the wheelspeed to produce an integer $+ \frac{1}{4}$ lobes using the chatter freq. f ($f/N_s = n + \frac{1}{4}$).

Work Regenerative Chatter

This type of chatter tends to occur at high workspeeds. In this case a small disturbance in the system causes a small transient wave to be ground into the workpiece. One work revolution later, this wavy work surface acts as a driving force to cause the system to vibrate again. Under unstable conditions a small wavelet can develop and extend around the work circumference.

Increasing the feedrate, or normal force, using larger and/or softer wheels, and reducing the workspeed prevent work regenerative chatter in the higher-frequency ranges. However, low-frequency structural modes of the machine are not inhibited.

If the work regenerative chatter cannot be suppressed, it may be possible to run the work speed at such a value as to cause an integer $+ \frac{1}{4}$ wavelengths per work revolution. That method is effective only if the number of waves per revolution is less than 10 approximately.

Another alternative is to reduce the cutting stiffness K_w. Since,

$$K_w = \frac{v_w W}{\Lambda_w} ,$$

lower work speed, narrower width of cut, and keeping the wheel sharp (high Λ_w) tends to eliminate the chatter. Increasing the static stiffness or the dampening in the offending structural mode also will tend to eliminate work regenerative chatter.

Wheel Mottle Patterns

These random patterns repeat at wheelspeed frequency but do not have a straight-line character. They are caused by local hardness and stiffness variations in the grinding wheel and are not the result of mechanical vibration. They are sometimes hardly measurable and appear only as a visual imperfection of the surface finish. Periodic roughness in a Talyrond chart is, at times, due to local hardness variations in the grinding wheel as shown by Hahn and Price.[7]

There is no complete cure for the patterns, but they can be suppressed to a certain degree. If the ratio of wheelspeed to workspeed is set to equal an integer + .5, the patterns will tend to be suppressed.

The patterns can be very prominent visually when the hard or stiff zone of the wheel operates in the "cutting region" (see wheelwork characteristic chart, Chapter 1), while the softer zone operates in the "ploughing region" since the surface finish produced in the 2 regions are significantly different. The cure, in this case, would be to change the finish feedrate or sparkout so that the grind terminates with all zones of the wheel operating completely in either the ploughing or cutting region but not straddling the "ploughing-cutting" transition.

References

1.   Hahn, R. S., "An Investigation of a Force-Adaptive, Creep-Feed Control for Improving the Rounding Capability of Flexible Grinding Systems," Proc. NSF, 12th Conf. on Production Research and Technology, SME, Dearborn, Mich.

2.   Hahn, R. S., "The Influence of Threshold Forces on Size, Roundness and Contour Errors in Precision Grinding," Annals of the C.I.R.P., vol. 30/1/1981, pp. 251-54, Hallwag Ltd, Berne, Switzerland.

3.   Hahn, R. S., "On the Universal Process Parameters Governing the Mutual Machining of Workpiece and Wheel Applied to the Creep-Feed Grinding Process," Annals of the C.I.R.P., vol. 33/1/1984, pp 189–192. Hallwag Ltd, Berne, Switzerland.

4.   Snoeys and Brown, D., "Dominating Parameters in Grinding Wheel and Workpiece Regenerative Chatter," Proc. 10th International M.T.D.R. Conf. 1969. pp. 325-348, Pergamon Press, Elmsford, N.Y.

5.   Inasaki, I., and Yonetsu, S., "Regenerative Chatter in Grinding." Proc. 18th International M.T.D.R. Conf. 1977, pp. 423-29, Pergamon Press, Elmsford, N. Y.

6.   Hahn, R. S., "Grinding Chatter in Precision Grinding Operations—Causes and Cures." SME Paper No. MR78-331. SME, Dearborn, Mich.

7.   Hahn, R. S., and Price, R. L. "A Nondestructive Method of Measuring Local Hardness Variations in Grinding Wheels," *Annals of C.I.R.P.*, vol XVI, pp. 19–30. Pergamon Press, 1968.

Index

M

Machining-Elasticity Number, 13
Metal-Removal Parameter
(see work-removal
parameter)

N

Normal Force Profiles, 173

P

Penetration Rate, 7
Ploughing-Cutting Transition, 8,
185
Ploughing Regime, 8, 185, 291
Power, 13, 235
(see also specific power)
Process Planning, 3

R

Reciprocating Grinding
effect of workspeed, 254
equations, 253
principles, 251, 283
surface integrity, 64
Residual Stress
(see surface integrity)

Rounding-Up Process, 176, 339
effect of threshold force, 180
Roundness Criterion, 181

S

Sensors, 339
Sizing
dresser diamond, 172
Sparkout, 56, 185
Specific Energy, 35
Specific Power, 13, 35, 52, 275
Stock-Removal Rate
cylindrical, 31
honing, 336
vertical spindle, 237
Stock Variations, 175
Surface Finish
coated abrasives, 278
determining factors, 347
effect of T_{ave}, 61
effect of wheelwear, 349
honing, 326
mottled patterns, 355
Surface Integrity
control of, 349
creep feed, 286, 290
cylindrical grinding, 65
honing, 303
shoulder grinding, 17
surface grinding, 255
Surface Profile, 114
System Rigidity, 171